Alison Mitchell

THE NEW PENGUIN GUIDE TO PERSONAL FINANCE

1992–3 Edition

Make Your Money Work for You

LONDON NEW YORK SYDNEY TORONTO

This edition published 1992
by BCA
by arrangement with Viking Books Ltd

First reprint 1993
Second reprint 1993

Typeset by DatIX International Limited, Bungay, Suffolk
Printed in England by Clays Ltd, St Ives plc

CN 4185

To my husband Ronald,
with love

Contents

Acknowledgements

Checking the detail of this book was a Herculean task and I am extremely grateful to all the specialists who gave their time and effort so freely. I am particularly indebted to Pauline Hedges of the British Bankers' Association for her help. I also benefited from the help of Jackie Manning, National Savings; Stuart Valentine, ProShare; Chris Anderson, Unit Trust Association; Fiona Monro, Association of Investment Trust Companies; estate agent Faron Sutaria; Sue Anderson, The Building Societies Association; Douglas Ferrans, Scottish Amicable; Aileen Kimber, Association of British Insurers; Sheila Longley, BUPA; Derek Thompson and George Marshall, Noble Lowndes; Paddy Ross, Leith Citizens Advice Bureau; school-fees specialist Stephen Whitehead; home-incomes plans specialist Cecil Hinton; and accountants Viscount Mackintosh of Halifax at Price Waterhouse and Emma Harding. The charts and figures came courtesy of Scottish Amicable, Nationwide building society, BUPA, and Noble Lownes.

But it is to my husband Ronald Pullen that I am most grateful. For his help and encouragement throughout and for the forbearance of my children Laura and Jamie, my thanks.

Alison Mitchell, Richmond, 1992

1 You and Your Money

Why you are reading this book · How to save yourself money

When I did the first-ever live television phone-in on money, on the second morning of BBC *Breakfast Time*, I was asked 'Where is the best place for my savings?'

Over the years questions from callers dialling into that famous red settee covered everything from animal insurance to zero-rating on VAT. But it was that first question that swamped all the others. 'Where is the best place for my savings?' It is a question I have been asked ever since. For, deep down, we all sincerely want to be rich. That is nothing to be ashamed of – we work hard for our money so why should we not want our money to work hard for us? Whether you are the sort of person who believes money was made round to go round, or flat to pile up, you will want to make sure that you are getting the best value for your money.

Just as there are always new ways to spend your money, there are new ways to save it too.

Fifteen years ago you would have gone to the building society for your mortgage and your bank for a loan. Now, you can go to either for your holiday traveller's cheques. In some respects, the two have become interchangeable, as both banks and building societies turn into financial supermarkets, often offering identical services.

In the last decade, four out of five of us have been getting better off. Because we have more money left over after we've fed and clothed ourselves this new-wave prosperity has been lapping away at the unyielding institutions in the City of London.

Until fairly recently few small investors chose the stock market as a natural haven for their spare cash. Now some 10 million private individuals – encouraged by the government's sell-off of state industries – have money invested directly in shares. That is 22 per cent of the adult population.

Surprisingly, there are as many shareholders as there are trade unionists in Britain today. As the joke goes, Marx and Engels have been replaced by Marks and Spencer.

It is easy to see how far-reaching the share-owning democracy has become. At one time it was only the quality newspapers that gave you the City news. Now tabloids are awash with financial columns.

The sale of British Telecom, which brought some 3 million new investors into the City, and later Sid and his gasmen colleagues, took dabbling in the market off the back burner. All of a sudden it became fashionable and profitable for the big City stockbrokers, banks and building societies to court the small investor and take on the little man again.

And the little lady. Women have traditionally held control of the purse strings and now, more and more, when it comes to budgeting, the fair sex are becoming the financial sex.

I began this chapter with the most common question I'm asked. A close second would be 'How can I pay less tax?'

But whatever your money queries, this book sets out to provide the answers, or to show you where you can easily find the answers for yourself. I have divided it into three parts – Savings and Investment, Tax, and Family Money. Each part takes you through the subjects from the easy-to-understand early stages right up to the complicated fine print of current legislation. Every chapter has a summary at the top, and an indication of other chapters on similar topics that you might like to read as well. Where you start – and where you finish – depends on how much you already know and how much you want to know about money.

Dip into the chapters or parts of chapters that are most relevant to your financial affairs. If you are interested in inheritance tax planning you probably won't need the advice on how to budget your monthly income. Though anyone starting with the simple chapters on savings and investment may soon graduate to the more complex ones! If there are other topics that are relevant to the part you are reading, I have detailed a cross-reference at the beginning of each chapter so that you can easily find them.

If you look after your own finances this book will help you to get the best deals.

If you already have a financial adviser you will learn how to ask the right questions and gauge the quality of his or her replies.

Only one person really has your financial interest at heart – and that person is you.

I guarantee that this book will save you money. You will get more interest, pay less in tax, get more for your house when you sell, or get a better deal for yourself on retirement.

I'm sure – indeed I guarantee – that if you take my advice you will get back more than you've paid for this book.

Whatever sex you are, whatever age you are, no matter the state of your bank account, this book will be able to help you.

PART ONE

SAVINGS AND INVESTMENT

2 First-step Savings Accounts

What's the best type of savings account to open first · How to choose the account for you · Types of savings accounts · Comparing interest rates · Easy access and rainy-day money

See also
- *Chapter 3, Building Your Savings*
- *Chapter 22, Your Personal Budget*

No matter how little money you have coming in each week, if you have an income at all you can, and should, be saving something.

The problem is that there is never anything left at the end of the week, or month, to put away for that rainy day. So nothing gets saved however good your resolutions were at the beginning of the year.

What you must do is save first. Use the budget planner in Chapter 22 to work out how much you think you can save. Be realistic. And as soon as you get your salary put that money into a separate account.

When I first started work, I put £10 a week into what I called my R D fund. (By that I meant my rainy-day fund.) I did it so that I would always be prepared for a rainy day, a financial emergency I hadn't foreseen. If I had a fit of pique at work, I could just leave my job and have a little money behind me to cushion the fall. In fact I never needed the fund for that, but I was surprised how quickly and painlessly that £10 a week grew into £520 a year, then a nice fat lump of interest was added, and suddenly I felt quite rich.

Of course you don't have to save as much as £10 a week if you can't afford it. How much you choose to save depends on your own personal financial circumstances. The key to succeeding without really trying is to put the cash away *every week* – in good times and in bad. If you can add more to it well and good, but never, ever, miss a payment.

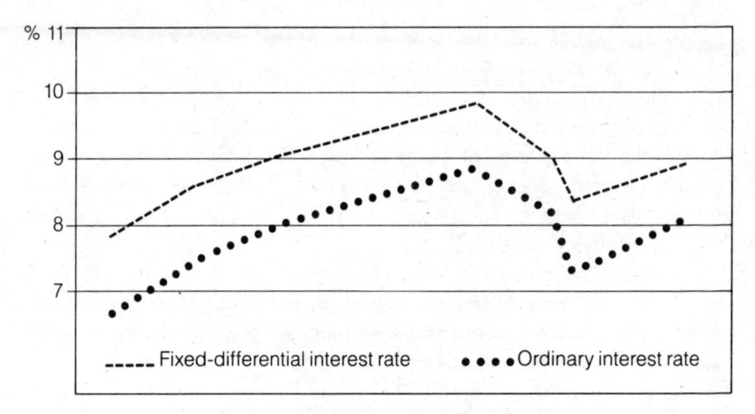

How a fixed-differential rate works

YOUR FIRST SAVINGS ACCOUNT

The first savings account you open should be a 'no-risk' account. No matter how much money you have, you should always have one 'no-risk' account for emergencies. After all there is no point in having all your millions tied up in a really good investment if you can't get your hands on the cash when the roof blows off your house.

In fact, there is no such thing as a completely risk-free investment. After all, it is just conceivable that all the banks and building societies could fail and you'd be left without your deposit. But that risk is so minute as not to be taken seriously, so when talking about no-risk accounts I am referring to High Street bank accounts, building society accounts and National Savings accounts.

What these institutions offer is interest on your money. The lump sum – the capital – that you deposit remains the same and is always there for you to withdraw if you want. In return for using your money to run their business, the banks, building societies and the government, who run the National Savings accounts, offer you interest.

The rate of interest depends on the prevailing economic climate and will go up and down along with all the other interest rates, such as the mortgage rate. As a general rule of thumb, the building societies tend to offer better basic savings accounts than the banks, and sometimes slightly better deals than the National Savings accounts do, too.

Some accounts offer fixed rates of interest. That means that before you make your deposit the bank, building society or National Savings will tell you the rate of interest on offer and that rate remains fixed throughout the life of the account. It can be one month, three months, a year, even, in the case of National Savings certificates, several years. You'll know in advance how long the deal is for. These accounts, often called bonds, can be worthwhile in times of high interest rates, but avoid them when rates are low.

A good halfway house is to go for what is known as 'fixed-differential' accounts. That means that, in return for agreeing to deposit a large lump sum, or tying up your capital for a certain length of time, you'll be offered a better rate of interest than other depositors. Say 1 or 1½ per cent more than the ordinary rate. And that differential, that 1 or 1½ per cent, will be fixed. So if interest rates in general go up or down, yours will too, but you will always get the set amount more than the ordinary rate.

HOW TO CHOOSE THE ACCOUNT FOR YOU

In olden days, saving was easy. If you chose to save, you chose a bank deposit account. No more. There are as many types of account on offer as there are types of saver. Which one you choose depends on your financial profile.

And to choose your first and most basic savings account, you really only need to answer three basic financial questions:

1 How much am I thinking of saving?
   (*a*) under £500 (*b*) over £500
2 How quickly will I need to get at the money?
   (*a*) daily (*b*) monthly
3 Am I a taxpayer?
   (*a*) yes (*b*) no

All the (*a*)s: You should have your cash in the most convenient building society ordinary share account.
All the (*b*)s: You should have your cash in the National Savings investment account.
(*a*)s and (*b*)s: You should read to the end of the chapter.

As a general rule, the more you have to save, and the longer you're prepared to tie it up, the higher the rate of interest you will get.

Size of lump sum

So if you have more than £500, even more than £100, check with your local building societies whether or not they offer an account which will give you extra interest on larger lump sums. Remember an extra 1 per cent on £500 a year is only worth £3.75 to you once the tax has been deducted from the interest, so don't go a bus ride away to get the deal. But you could use a building society postal account.

Notice of withdrawal

You can also get a better rate of interest if you are prepared to give a week, a month or even three months' notice of withdrawals. So it might pay you to keep say £500 in an ordinary building society account and the rest of your emergency money in a ninety-day high interest account. That way you have some money that is accessible, and the rest is earning nice high interest for you.

Tax

Your tax position is crucial in determining where you put your money. If you are an ordinary rate income-tax payer, you might as well put your lump sum into a bank or building society account. Tax is deducted from the interest before you get it, so you don't have to worry about paying it later. Though you do have to declare your interest to the Inland Revenue on your tax return if you complete one. See Chapter 7, Understanding Income Tax.

Non-taxpayers can claim back any tax they have paid on the interest. Fill in a form at your local bank or building society branch before the start of the tax year and the interest will be paid gross. Or write to the Inland Revenue to reclaim any tax already paid. Or choose instead one of the National Savings accounts that pays interest gross, that is without deducting tax first. The accounts that pay interest gross are:

- investment account
- capital bonds
- income bonds.

Taxpayers can also opt for these accounts, to take advantage of the higher rates of interest, but they'll have to declare their savings to the Inland Revenue and pay tax on the interest.

Which is the best emergency account for you?

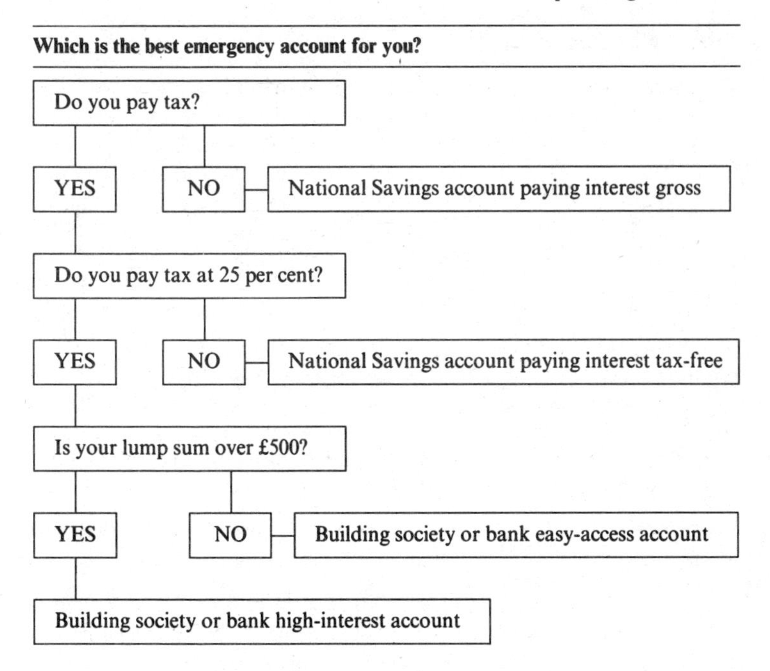

High-rate taxpayers should also avoid the ordinary building society and bank accounts. They have to pay tax on the interest at their highest rate of tax, that is 40 per cent, so will be asked for further tax on all the interest they get.

They should choose instead National Savings certificates, which offer interest tax-free. That means you don't pay tax on the interest at all, so they're worth a lot to high-rate taxpayers.

The National Savings accounts offering tax-free interest are:

- National Savings certificates
- Index-linked certificates
- yearly plan.

Non-taxpayers should avoid these accounts completely.

TYPES OF SAVINGS ACCOUNTS

Bank accounts

Seven-day deposit accounts. These are ordinary bank accounts which offer interest on deposits but deduct a week's interest if you don't give a week's notice of withdrawals. In Scotland, seven-day deposit accounts are rare animals. Their basic bank accounts are simple deposit accounts, though they do tend to have ATM (hole-in-the-wall) cards as an additional facility. The rate of interest tends to be the same as in England, but you don't have to give seven days' notice of withdrawals or lose seven days' interest.

Fixed-rate account. These tend to be for one month, three months or a year, and offer fixed rates of interest.

Investment account. These pay a higher rate of interest, but you will have to tie your money up for longer or invest a lump sum of over £1,000.

Monthly-income account. You'll need a lump sum of over £1,000 but you'll get a good rate of interest and the interest is paid monthly.

High-interest cheque account. Usually for deposits of over £1,000. You will get all the extras normally associated with a cheque account – cheque-book, ATM card, standing orders, direct debits – and interest on your money as well. The larger the balance you have, the more interest you will get. But rates are lower than investment accounts, so don't put all your savings here.

Cheque account. Most banks now offer interest on current accounts. You won't get quite as much as you would if the cash were in a savings account, but it is usually a decent percentage. With most of the banks the interest rate is tiered. The more money you have in the account, the higher the rate of interest you will get.

However, look closely at the small print before deciding which account to go for if you want the best deal. The tiers on higher rates of interest vary from bank to bank. Some charge management fees on the account; others allow a charge-free overdraft up to a certain level. Choose the account that suits your financial circumstances best.

Building society accounts

Share account. This is the ordinary account. You can pay in as much as you like as often as you like, and withdraw whenever you fancy it –

even in the middle of the night if your building society has an ATM (automated teller machine – hole-in-the-wall machine).

Instant access account. This is now more popular than the ordinary share account. It works in the same way, offering instant access, but has some advantages. Tiered interest means that when your savings pass certain levels – say, £500 or £1,000 – the rate of interest automatically increases. If you withdraw below the set levels, the rate of interest automatically decreases.

Seven-day, thirty-day and ninety-day accounts. These are for people prepared to give notice. The more notice you agree to give, the higher the rate of interest you'll get. Some building societies will allow instant access on these accounts, but with the appropriate loss of interest in lieu of notice, others insist that you give the notice. Check the fine print before you sign up.

Monthly-income account. You'll need a lump sum of over £1,000 but you'll get a good rate of interest, and the real plus factor for many people, particularly pensioners, is that the interest is paid monthly.

Regular saving

Banks and building societies also offer accounts for people wanting to save regularly, month by month. Because the cash is committed the rate of interest is higher than for ordinary accounts, but you are normally limited to one withdrawal every six months. These are ideal accounts for your rainy-day money, but are thinner on the ground now than they used to be.

TESSAs

Most banks and building societies also offer a TESSA – that is a Tax Exempt Special Savings Account.

This is a savings account, with particular rules, which pays interest tax free and is a must for all taxpayers (over eighteen) who can tie their money up for five years.

The rules are very simple:

- this is a five-year account
- you can pay in up to £9,000 – £3,000 in the first year, £1,800 in the next three years and £600 in the last year

- interest is tax free
- if you withdraw the interest before the five years is up you have to pay the tax on it
- if you withdraw more than the interest the account is closed.

Banks and building societies which operate these accounts set their own rate of interest on them – so go for one offering a top rate. But watch out for charges, penalties for switching your TESSA to another bank or building society, or the need to set up a 'feeder' account to lodge your TESSA savings for subsequent years.

National Savings accounts

Investment account. High-interest savings account paying interest gross but you have to give a month's notice of withdrawals. The interest is taxable. You can invest £5 to £25,000.

Ordinary account. Savings account offering more interest on savings of over £500 but still the poor relation of the NS accounts. Good for high-rate taxpayers as the first £70 of interest is tax-free. You can invest £5 to £10,000.

Income bonds. For lump sums of over £2,000 the interest, paid gross (though taxable), is sent to holders monthly. Three months' notice of withdrawals is required. Further deposits in £1,000 tranches, up to a maximum of £50,000.

Capital bonds. This is a five-year home for lump sums of over £100. The interest, which increases every year you hold the bond, is added to the lump sum. The interest is taxable and must be declared on your tax form, so, if you are a taxpayer, you will find that you have to pay tax on interest that has not yet been paid to you. When Capital bonds were first launched the interest started at 5.5 per cent in the first year, rising to a massive 20.6 per cent in the last year, and once you buy you are guaranteed these rates of interest. However, the rates change as interest rates go up and down (although each series has a guaranteed fixed rate), so buy Capital bonds only at times of high interest rates. Capital bonds are ideal for non-taxpayers who don't mind tying their money up for five years.

Savings certificates. Accounts which offer a fixed rate of interest tax-free. Interest is paid on a sliding scale, with the maximum being given to savers holding for the full five years. If you want your money back you will generally get it in just over a week after you ask for it. Savings certificates are only worth buying when interest rates generally are high because you continue to benefit from the high rates for five years and, usually, on that time-scale, interest rates that are not fixed have fallen. If you are transferring money from a matured issue to the current one, the maximum holding is £10,000.

Index-linked certificates. Like savings certificates except that the interest is twofold. Every year the rate of interest that is added equals the rate of inflation, plus a tax-free bonus if you hold for the full five years.

Tip

If the tenth anniversary of your first or second issue index-linked certificates is approaching, try not to sell. Hold on until the date is passed, and you will qualify for an extra 4 per cent interest.

Yearly plan. This account works as a monthly savings plan for the first year. The cash is then invested for a further four years at a fixed rate of tax-free interest. You can invest between £20 and £200 a month and if you want your money back you should get it within fourteen working days of your notice of withdrawal.

Children's bonus bond. This was launched by National Savings to attract children's money. It is a five-year hold, with a rate of interest paid annually and a bonus added at the end if you hold for the full term. You can buy the bond in £25 chunks – but you can't breach the upper limit of £1,000. Although Inland Revenue rules prevent parents from giving their children money that will produce an income of more than £100 a year, the children's bond is exempt from this because the interest is tax free.

FIRST option bond. Introduced in the 1992 Budget, this bond is designed to appeal to basic-rate taxpayers. The rate of interest will be fixed, tax paid at 25 per cent and guaranteed for a year from the day that you buy it. Because it is an annual bond it is better to leave your money in for twelve months. Two weeks before the anniversary date, National Savings will write to you, telling you the new rate of interest and asking if you want to leave your money where it is. If you do nothing, you automatically lock yourself in for another year; otherwise you have a two-week window to withdraw some or all of your money or just the interest if that is what you want. You have to invest over £1,000 and less than £250,000. No notice is necessary for withdrawals, but you lose a lot of interest if you don't hold for twelve months.

Premium Bonds

Although Premium Bonds are sold by National Savings, they are not really an investment as there is no guarantee of any return at all on your money. In fact the chances of Ernie paying you the monthly jackpot of £250,000 are 1 in 1.9 billion. Slight to say the least. But the chances of your one Premium Bond paying out any prize at all are better, at 1 in 11,000. Total Premium Bond prize money is calculated using a notional interest rate, lower than elsewhere, but it is always worthwhile tucking a few bonds away – after all someone wins the jackpot every month.

There are plenty of other safe, interest-bearing homes for your cash but I think everyone should have at least one account from the above list before going on to more sophisticated options.

Other savings accounts

High-interest cheque account

One way of making your savings work harder is to deposit cash in a high-interest cheque account. Exactly as the name implies, this type of account offers you the facilities of a cheque account, with the high interest of a deposit account.

Of course there are restrictions. You have to have a large lump sum, mostly over £1,000 or even £2,500, and you can write only large cheques, usually over £250 or £500. This is to stop you using the account for all sorts of small transactions which would be expensive for the companies to operate. But it is ideal if you're keeping your holiday savings there. You can build up the lump sum, see it earning nice high interest, and use it to pay the travel agent's invoice, for example. Most High Street banks and several City institutions offer these accounts.

Local authority loans and bonds

Local authority loans and bonds used to be an option for smaller savers. This is no longer so. Although you can still lend money to a local authority and get interest on your loan, you would have to do a lot of shopping around yourself to find the local authority paying the highest rate of interest.

If you want to invest more than £100,000, there is a company that will do this work for you – Sterling Brokers. Tel. 071 407 2767.

You can also buy local authority bonds that are traded on the stock market. However, the yield (that is, the interest that they pay you) is not as high as you can get elsewhere, so I would give these a miss.

Friendly societies

Friendly societies are a must for a small portion of your savings if tax efficiency is your main priority. They started up in the last century to help people save for their burials and, as an incentive, the interest on the savings was allowed to accumulate free of all tax.

As a result smart investors began to use friendly society savings accounts for a larger proportion of their savings, so the tax loophole was then closed. The amount of money you are allowed to save in a friendly society account (and you can have only one account) has been strictly limited to £150 a year (£12.50 a month) and there is usually a minimum of around half these amounts. Investors have to be over eighteen to open a policy in their own right.

The money is invested, rather than deposited in an interest-bearing account, but few risks are taken with the cash. You should do much better than leaving the money in a building society account, but perhaps not as well as if you bought an ordinary endowment policy which might invest in slightly riskier shares.

Friendly society contracts run for a minimum of ten years and you can't get your tax-free return out until it has been going for at least seven and a half years. But like ordinary life insurance policies the surrender value in the early years is very poor indeed. So if you can't keep the policy going for ten years, don't start in the first place.

To encourage you to invest more, many societies offer hybrid schemes. If you invest more than the friendly society maximum, then the extra part is invested in a unit trust or managed fund.

COMPARING INTEREST RATES

Working out where to get the best savings rate is easier than it might seem despite the variety of schemes on offer. Many of the national Saturday and Sunday papers regularly print tables showing who is offering what, and the penalties for withdrawals. So all you do is choose the one that suits you best. Much easier than trying to find out all the rates for yourself. But remember to differentiate between accounts which offer interest gross, that is without deducting tax from the interest, and those offering interest net. The latter may seem to be lower, but if you are a taxpayer, and will have to pay tax on the interest anyway, you might be better off going for a net account.

EASY-ACCESS AND RAINY-DAY MONEY

I'm not a fan of having too many similar building society or bank accounts. For one thing it is a nuisance to keep track of them and you have to fill in all the interest payments on your tax return, if you get one. For another, the more you have in one account, the higher the interest you're likely to get. So by spreading your savings around, you could be costing yourself money.

But it is often to your advantage to have two accounts. One for your easy-access cash and one for your rainy-day money.

Easy access cash. The first, with as little as possible in it for emergencies, should be convenient – that should be your main criterion. It should either be near where you work, near where you live or near where you shop. And do yourself a favour by choosing a building society that isn't always too busy – if it has a hole-in-the-wall machine outside that you can use instead of joining the queue inside, so much the better.

But don't have too little cash in the account. Building societies are awash with people having less than £100 in their accounts and often popping in to withdraw or deposit. It is costing them too much money to operate these accounts and they are starting to pass these charges on. Keep an eye on your building society and keep above the limits if it imposes charges.

Rainy day money. Your other account should be chosen for interest. The higher the interest the better it is for you. And you can probably afford to tie up your money a bit, thus getting better interest, on the basis that your ready cash is in the easy-access account.

You can operate building society accounts by post. Often the societies offering the highest rates of interest are small societies who don't have the heavy costs of running a lot of branches. If there isn't one in your High Street you can deal with a branch elsewhere by post or phone. Again, you'll find tables in the financial press highlighting the best deals.

3 Building Your Savings

Income versus capital growth · Unit trusts · Investment trusts · Gilts

See also
- *Chapter 4, Buying and Selling Shares*
- *Chapter 10, Tax and Your Investments*

For the moneywise the second tranche of savings is almost more important than the first. This is the cash they will be relying on for their future.

The first savings account that you open tends to be for emergencies, and for short-term savings – holidays, a car, or for just about anything that might come up. But the next step up is the savings for a deposit on your first house, for your old age or for tiding you through any unforeseen bad times. It is cash that you want to see worth something in the years to come, that will work hard for you after you have worked hard for it.

And the first thing to do is to make a very simple choice. Do you want income from this money, or capital growth?

INCOME VERSUS CAPITAL GROWTH

Income. If you can't live on the money you have coming in every week then you need income from your lump sum. That means it will have to be invested at the best possible interest rate to provide you with a regular addition to your budget.

Capital growth. If you can live on the money you have coming in, then you want capital growth from your lump sum. That means it will not give you any additional income at the moment but the lump sum itself will grow and become worth more over the years. At the very worst it should keep pace with inflation – but a money manager worth his salt should be able to do much better than that.

You can opt for schemes which give you both income and capital growth but, obviously, a little of both means not as much of either.

You can get income from your savings by investing in a bank, building society or National Savings account but, if you are prepared to take a slight risk with your lump sum, you should be able to get a better deal.

And here you come up against one of the most basic of all money rules. To get the best deals for your cash, whether you want income or capital growth, you have to be prepared to take some sort of risk. The higher the risk the higher the potential reward. But think very carefully – taking a risk means that you can actually lose part or all of your savings. If you can't afford or don't want to do that, don't choose a risky investment.

To maximize your return, whether you are looking for income or capital gain, you really have to link your money into the stock market. There are two ways of making money on the Stock Exchange. One is to buy shares that pay out a good and rising dividend every six months – that should give you a better income than you'd get from a bank account. The other is to try to choose a company whose share price is on the way up, and that would give you capital growth.

But share prices can fall far and fast too. In the autumn of 1987 the *FT*-SE 100-Share Index, the chart which measures share rises and falls, slumped by 30 per cent in two months. Triggered by a panic on Wall Street, share prices had their worst-ever one-day fall here on what was to become known as Black Monday. Many big investors were very nearly wiped out. So, no matter how high share prices seem to be going, don't be lulled into a false sense of security. They can, and do, fall very sharply.

To minimize your risk let the professionals spot the potential winners. If you've a large lump sum you can get advice from a stockbroker and invest directly in shares, otherwise opt instead for a unit or investment trust. And I wouldn't go directly into shares unless you have over £3,000 to invest. Less than that and you should stick to unit and investment trusts.

Unit and investment trusts are really just funds of investors' money professionally managed and invested in a wide spread of shares. Every investor has a share of the whole so that his or her risk is spread much more widely than would be possible with your own small lump sum. For example, if you invest £500 in a unit or investment trust it will, along with everyone else's money, be used to buy shares in dozens of companies.

If you invested your £500 directly in the stock market yourself you'd only be able to buy a few shares in one company so your risk would be much greater. You would be tied to the fortunes of that one company, rather than spreading your risk over dozens.

UNIT TRUSTS

Among small investors, unit trusts are the most popular way of investing in the stock market at the moment. And they're very flexible. You can invest as little as £250 (though many companies ask for a minimum investment of £500 or £1,000, or £20 a month with a regular savings plan) or as much as you want, regularly or occasionally, with one or with as many of the unit trust groups as you fancy. And there are around 160 groups managing around £57 billion in 1,422 funds.

If you want to invest you can just choose the unit trust or trusts you like the look of and send off your cheque to the management group.

The cash will then be invested in shares by professional fund managers and their fee is included in the price you've paid for your units. The more people who want to join the trust, the larger it gets.

And you can get your money back at any time by selling your units.

There is one major catch. At any one time, the price you get when you are selling your units is a lot lower – usually between 5 and 7 per cent – than the price you would pay to buy.

This difference is called the 'spread', and it is the money the management company takes to cover its costs and make its profit. There is also an annual management charge of between ¾ of a per cent and 1½ per cent of the value of the fund. And that is deducted from the income before any dividends are sent out to unit holders.

In order to make any money at all you have to see the price of the units rise by 7 per cent at least. So don't think of unit trusts as a short-term investment. If you are not prepared to put your money in and leave it for at least three years, don't buy unit trusts.

And of course the price of the units can and does go down. You could invest £1,000 and sit back and see the price drift down through £900, £800 and even further. Which is another reason why you should see unit trusts as a long-term hold. It would be a bad fund manager who couldn't at the very least maintain the price of your units over three or more years, unless there was a long-term fall in share prices generally or he was investing in a sector or area of the world that hit a downturn.

However, you don't have to hold for that length of time. If you find that the price of your units goes up and you want to take the profit, there is nothing to stop you from so doing.

How to buy unit trusts

Buying unit trusts is very easy. There are three main ways:

1 through an intermediary such as a financial adviser or bank or a unit trust salesman attached to a particular group of funds
2 by cutting out a coupon in a newspaper or magazine
3 by phoning or writing to the unit trust company directly.

Whichever way you choose, the result and the costs to you will be the same. If you use an intermediary he will do all the paper work for you and deal with the unit trust company – and unless you are paying him a fee yourself (which would be unusual) it won't cost you anything. He gets his commission from the unit trust company.

If you cut out a coupon, it is usually for the launch or relaunch of a fund and there's often a special offer, such as a discounted price or extra units if you get in quickly, or a fixed-offer period when the price of the units won't change for, say, a month. All you do is fill in the coupon, attach your cheque and send it off. Usually the minimum investment at this time is £500 or £1,000.

If you want to deal with the company yourself, all you do is phone up the unit trust managers. You can write but you could find that the price of the units changes between when you post the letter and when they receive it. When you phone up, ask the managers the buying price of the units (it is known as the 'offer' price) and, if the price suits you and you want to deal, tell them how many units you want to buy.

And that's it – you are committed, and so are the unit trust managers. You then have to send off your cheque. Seven to ten days later you will get a contract note and about a month after that your unit trust certificate will arrive. Keep it safe; if you lose it you'll spend quite a bit of time, effort and money replacing it.

Incidentally, dealing directly with the unit trust company won't necessarily save you any money. The commission that they save by your not dealing through a commission-based salesman is not passed on to you; it is kept by the unit trust managers.

The price of the units

Unit trust prices vary according to the value of the underlying shares. Exactly how much you have to pay depends on the system of pricing that the unit trust group uses.

There are three main options:

- historic pricing
- forward pricing
- combination of the two.

Historic pricing. You can buy or sell units at the price given by the most recent valuation. Most unit trusts are valued at the same time every working day, but different groups will choose a different time. If your unit trust group uses this method you will know exactly what the units will cost you, or how much you will get for the ones you are selling.

Forward pricing. You buy or sell units at the price worked out on the next valuation, so you don't know when you place your buy or sell order precisely how much you will have to pay or how much you will receive.

Combination. Some unit trust groups which value their funds during the day at, say, 11 a.m., will use forward pricing from the start of business until 11 a.m. then change to historic pricing for the remainder of the day.

If your unit trust group deals on a historic pricing basis it still has to offer forward pricing if the customer asks for it, or if the stock market moves more than 2 per cent up or down since the last valuation.

How to sell unit trusts

There are two main ways to sell unit trusts:

1 through your financial intermediary;
2 by phoning or writing to the unit trust company directly.

If you have a financial intermediary just tell her what you want to sell and she will do all the work for you.

If you want to deal directly with the company, just phone up the unit trust managers – again it is more immediate than writing – ask the selling price (it is known as the 'bid' price) and, if it suits you, tell them how many units you want to sell. You don't have to sell all your units at the same time; you can deal in any number you like (provided you own them) so long as you leave the minimum investment in the fund. Different funds have different minimum investments, but it is usually £500 or £1,000.

Your cheque from the unit trust company will arrive seven to ten days later. So it is quite easy to take some of your profits without selling the entire holding.

If you lose your certificate

Tell the company immediately if you lose or mislay your unit trust certificate. They won't just issue you with a replacement. You will have to go through a procedure of 'indemnification' – the company will tell you exactly what you have to do. It takes a week or so, and will cost you between £10 and £15, and you won't be able to sell any units in the meantime. So keep your certificate safe.

How to choose a unit trust

Unit trusts come in all shapes and forms. You can invest in general funds – those which tend to buy a wide spread of good solid British public companies – or specialist funds which will go for a geographical area such as the Pacific, Japan and the Far East, America and so on, or funds which put their money into property shares, or gold shares, or oil shares or into shares in companies that they think will recover from a current bad period (they are known as 'recovery' funds). The more specialist the fund the greater the risk, because all your eggs are in one basket, but the higher the potential reward.

You can choose a fund specifically for income (a high-yielding unit trust) or specifically for capital growth (with very little income) depending on what you are looking for, or you could go for one that offers both. Pick the type that best suits your finances.

Most unit trusts pay a dividend twice a year – unless you opt for one specializing in capital growth – but you can choose one that will pay out monthly or pay out a specific sum regularly to supplement your income. Income can be paid in one of two ways. The unit trust will either send you a cheque (from which basic rate tax will have been deducted) or it will offer you accumulation units – that means the dividends are automatically reinvested in the fund.

It would be impossible for me to tell you which unit trust to buy. I don't know which ones are going to perform best next year, and I don't know how much risk you want to take. But if past performance is any guide to the future you could opt for a unit trust that has done well in the past. They may not be the best performers over the next few years, but I doubt if they'll be the worst.

How to follow the fortunes of your unit trust

If you want to check how well – or badly – your unit trust is doing you'll need to know the price of the units. You can either ring the unit trust managers and ask them or, if your unit trust is not too specialized, look in the share-price columns of the *Financial Times* or the quality daily papers. It may take you a while to find it at first, but it will always be in the same place, so once you've found it you are home and dry.

All you do to work out the value of your holding is multiply the 'bid' price (the lower of the two prices) by the number of units you hold. But remember, by the time you come to sell, the price may have changed.

You will also receive half-yearly reports from the unit trust fund managers outlining how well – or badly – the trust has performed over the past half year, and what direction it is going to take in the following six months. If you disagree with their investment policy, sell the units.

More help

If you want a list of all the unit trust companies and their funds, or a helpful little booklet giving more detail on how to buy and sell unit trusts, write to the Unit Trust Association Information Unit, 65 Kingsway, London, WC2B 6TD (tel. 071 831 0898).

INVESTMENT TRUSTS

In many ways investment trusts are like unit trusts – they take your money, and that of hundreds of other investors, and use it to buy a wide spread of shares. That way, again, you spread your risk.

But investment trusts, in my opinion, are for the slightly more sophisticated investor – you tend to need a stockbroker to buy and sell them. But don't be put off by that – they are fairly simple to understand

and there's a good profit to be made out of investment trusts, so it is very worthwhile taking the time.

Over the past five years investment trusts have performed as well as unit trusts, over ten years they have performed much better.

It is reckoned that if you had invested £100 in an 'average' investment trust in 1981, it would have been worth £147 five years later; in an 'average' unit trust it would have been worth £131. By the end of 1991 your original £100 would have grown to almost £528 in an investment trust, just over £386 in a unit trust.

Of course, there is no such thing as an 'average' fund. Everything depends on the investment performance of the fund managers. But the figures do point to investment trusts in general outperforming unit trusts in general.

And investment trusts do have some major advantages. Investment trusts tend to have lower management charges and they don't have that front-end loading – the 5–7 per cent premium you pay when you buy unit trusts. So if a unit trust and an investment trust showed exactly the same growth over the years, the investor in the investment trust would do better because of the lower charges.

What is an investment trust?

An investment trust is a public company and its shares are dealt on the stock market just like those of any other public company. The only difference is that it doesn't manufacture widgets or sell pints of beer, it makes its money investing in other companies. Professional fund managers invest the fund's money in companies they think will do well, and if they invest wisely the price of the shares in the investment trust goes up – making you, the shareholder, a profit. An investment trust manager can also borrow money for the fund – something a unit trust manager cannot do – so that can help to make an investment trust more profitable than a unit trust.

But unlike unit trusts, an investment trust cannot grow once it has been set up. If you want to buy shares in an investment trust, you will need to find someone who is wanting to sell. Of course *you* don't have to find the seller, that is the function of the Stock Exchange, so you deal in the investment trust through a stockbroker, or through your bank. (See Chapter 4, Buying and Selling Shares, p. 32.) Increasingly nowadays, you can buy and sell investment trusts by sending money directly to the company managing the trust. Your money will still go through a broker to buy the shares in the investment trust, but the management company will get a better deal because they can offer the broker a lot of other business. So it will cut your costs.

Otherwise you can buy and sell the shares in the investment trust through a stockbroker, bank or building society.

Which investment trust should you buy?

Like unit trusts, investment trusts are either general or specialist, designed for income or capital growth, and it is up to you which you opt for.

But the price of the shares goes down as well as up so there is the same risk involved in buying them.

Choosing an investment trust is every bit as difficult as choosing a unit trust, except that there are a lot fewer of them. Around 260 at the moment. Once you've decided whether you want income or capital growth, a general fund or a slightly riskier specialist fund – and if so which country or commodity you want to invest in – you still have to choose a management company. Look at past performance and go with a company that has a good track record.

More help

A full list of investment trust companies can be obtained from the Association of Investment Trust Companies, Park House, 16 Finsbury Circus, London EC2M 7JJ (tel. 071 588 5347).

GILTS

If you're a little nervous of the risk involved in investing in the stock market you could hedge your bets by buying gilts instead.

Again it is a slightly complicated concept to understand, but worthwhile because you stand to make quite a bit of money – tax-free – out of them. Gilts, or, to give them their full name, gilt-edged securities, are in fact a loan from you to the government. Every year when the Chancellor of the Exchequer announces how he is going to raise the taxes he needs through the Budget, he usually leaves a shortfall. He will raise less in taxes than he is going to spend. And that shortfall is partly made up by the government borrowing money – in a small way through National Savings but in the main through selling gilts.

What they are offering are, in effect, IOUs. They will take your money, in tranches of £100, and in due course they will repay it. In the meantime they will pay you a fixed rate of interest on it, every six months. It is that simple.

In practice, dealing in gilts effectively and profitably is more sophisticated.

Dealing in gilts

Gilts are dealt on the stock market rather like shares. You can buy them and sell them whenever you like and the price of the gilt is governed by the interest rate it offers, and current interest rates. If the gilt was launched and the interest rate fixed at a time when rates were high, and a year or so later interest rates generally have fallen, the price of the gilt will rise. That is because investors can get a better rate of interest on the gilt than in, say, a building society account, so more people want to buy the gilt and that of course puts the price up. If a gilt was launched when rates were low and they have since risen, then of course the price of the gilt will fall because investors can get a better rate of interest on their money elsewhere.

Prices of gilts go up and down and they are quoted in fractions of a pound so you could see your gilt priced at say 93½ – that means the price for £100 is £93.50.

Example

If you wanted to buy £1,000 of gilts that were currently priced at 93½:
you could buy 1,000 × 93½ and that would cost you £935.00
or
you could buy 1,100 × 93½ and that would cost you £1,028.50.
On top of that, there would be dealing costs.

Dealing costs

To buy £1,100 gilts at £93.50 per £100:

£1,100 × 93.5	£1,028.50
Commission*	£29.50
Bargain charge	£3.00
Total bill	£1,061.00
Dealing costs	£32.50

* There is no VAT on stockbrokers' commissions.

Bargain charge. Around half of all stockbrokers ask you to pay a bargain charge – that is a set fee for every deal you do. The cost varies from around £1 all the way up to £25 and is expected to pay for 'compliance', the cost of policing the stockbroking system. It is worth asking, before you strike a deal, what the bargain charge will be. After all, a large charge on a small deal might wipe out a lot of your profits.

Of course you don't have to *deal* in gilts at all. You could buy, get the interest six monthly, and hold the gilt until the date on which the government have promised to repay the money – that's known as the 'redemption date'.

By doing that the investor knows how much interest she'll get over the years – and it can be anything from one year to thirty years – and the date she'll get her money back. On the redemption date, the investor will get £100 for every £100 worth of stock that she holds, regardless of what she paid for it.

You don't have to pay capital gains tax on any profit you make on gilts.

Using gilts in this way can help pensioners to manage their money for their old age, or parents and grandparents to fund a school fees plan. You know exactly how much your lump sum will be worth in the year that it matures and how much of a capital gain you are going to get on your lump sum. And there is no risk to that investment; the money is guaranteed by the government.

However, on top of that there is the chance that during the life of that gilt it will shoot up in price. In which case you can sell and take your profit. So you get the chance of a profit without any of the risk – not something you come across very often in the world of money management.

Checking the price

If you want to check the price of your gilt you can look for it in your paper. The *Financial Times* carries all the prices and the quality daily papers carry a selection of the more popular gilts.

Your gilt will probably be called either 'Treasury' or 'Exchequer' – the names are really meaningless.

Then will come the rate of interest paid on it – anything from 2 to 15½ per cent.

Next comes the date – 1992–2024 – that tells you when it will be redeemed.

So your gilt could be called Treasury 10pc 1995. That means that it will pay 10 per cent interest a year and will be redeemed at £100 in 1995.

Then you will see the price quoted – say £103.

After that come two things which are slightly more complicated:

1 *The interest yield.* This works out what the yield is on the gilt at the price you would have to pay for it. On our Treasury 10pc 1995 at £103, the yield will be less than 10 per cent because you paid more than £100 for it. In fact, the yield is 9.70 per cent.
2 *The redemption yield.* This takes into account the price you would pay, and how long the gilt has to run before it is redeemed. On the same example, if you paid £103 for the Treasury 10pc 1995, the yield to redemption would be 9.2 per cent.

Buying and selling gilts

If you want to put your money into gilts the easiest way is through a stockbroker or bank. You'll be charged a commission on the deal.

Rates are no longer fixed but there will probably be a minimum charge of £20 or so, and then costs will probably be around 0.5 per cent of your investment.

However, there is a much cheaper way into gilts – through the National Savings Stock Register.

There are two benefits here.

1 The six-monthly interest payments are made gross so if you are a non-taxpayer you don't have to claim the tax back because it is not deducted in the first place. Though, if you are a taxpayer, you'll have to pay income tax on the interest at your top rate.
2 Commission charges are much lower.

National Savings Stock Register commission rates

To buy gilts	
Up to £250	£1
Over £250	£1 + 50p per £125 of gilts
To sell gilts	
Under £100	10p per £10
£100 to £250	£1
Over £250	£1 + 50p per £125 of gilts

Although you now have a choice of sixty gilts, the drawback of dealing this way is that you have to do your transactions by using a special form available at your local large post office. This means that if interest rates change suddenly, there may be a sharp move in the price of gilts between your posting the form and the National Savings office receiving it. That could be good or bad news for you depending on which way interest rates moved, but it is a risk.

If you are buying gilts because you intend holding them to redemption, or for many years, use the National Savings Stock Register. A variation of a few pence between the price you expected to pay and the price you actually pay will make little difference to you. The reduced commission will.

4 Buying and Selling Shares

All you need to know about how to buy and sell shares · Choosing a winner · Keeping track of your investments

See also
- *Chapter 3, Building Your Savings*
- *Chapter 10, Tax and Your Investments*

Buying shares can be great fun. There is nothing better in the world of money than watching the price of the shares you bought go up and up. For the first time you are making real money without having worked for it. You will suddenly find that you check your share prices in the paper before looking to see what's on TV, and you'll sail perilously close to boring all your friends with day-by-day updates on your profits.

But there is a great cold turkey cure for new share-buying enthusiasts. And that is seeing your precious investments fall dramatically.

You'd be a fortunate person indeed never to buy a dud. And there is nothing like shouldering a substantial loss on one of your shares to drive home the message that buying shares is a risk investment.

Unlike depositing cash in a bank or building society account, where you know that the lump sum you paid in will always be there, share values can and do fall below the amount you paid for them.

Often it is not the fault of the company at all. In the autumn of 1987, major American investors decided that the US economy was on the skids so they sold shares. Overseas investors countered by selling the dollar. Stock markets around the world – including our own – went into their first major decline in more than a decade and over 30 per cent was wiped off share values by the end of the year. This despite the fact that our own economy was still sound, unemployment falling and output growing.

But it can also be a setback in the company you have invested in that knocks the share price. If the company gets into serious financial trouble, you could lose your money altogether.

Over 4 million people in this country bought shares for the first time from the government during the privatization campaign. Attracted by multimillion pound advertising campaigns they invested in British Telecom, British Gas and British Airways, added in TSB and found themselves sitting on mini-portfolios that were looking very healthy indeed.

Every single investor who bought these new issues made money on the deal at the time. Not everyone is so lucky.

If you don't want to buy shares yourself, there are less direct ways into the equity market. You could buy unit or investment trusts. Your money will be invested in shares – but you won't have to make the investment decisions on which shares to buy. Full details on unit and investment trusts are in Chapter 3, Building Your Savings.

And many people also invest in the shares of the company they work for, through share option schemes, or have money in personal equity plans. These subjects have a chapter of their own, Chapter 5, PEPs and Share Option Schemes.

THE STOCK EXCHANGE

The Stock Exchange started over two hundred years ago to bring together companies wanting to raise money for their business and individuals with cash to invest. It still performs the same function in raising new money for companies by selling shares to outside investors, but its more important role is as a market-place for investors to trade their shares.

In recent years, much of the business was taken over by the big institutional investors dealing in multimillions of pounds. They are the City institutions, such as insurance companies and merchant banks, who look after the £150 billion or so of funds paid into pension schemes and insurance plans and who buy and sell hundreds of thousands of shares at a time.

In the last ten years, the small private client has made a comeback. The crash of 1974 hit small investors badly and it took almost a decade to tempt them back.

Spurred on by the government sell-off of state industries, and buoyant share prices, the smaller investor is a lot less rare now than he used to be. It is reckoned that almost one in five adults owns shares.

To help new, and not so new, investors the Stock Exchange has set up an Investors' Club. It is now run by ProShare, an independent body set up to promote share ownership. There is an annual subscription which pays for newsletters, information on investors' protection and a guide to which stockbrokers encourage the small investor. You can get further information from ProShare at 13–14 Basinghall Street, London EC2V 5BQ (tel. 071 600 0984).

But how much money do you need to be a 'small investor'?

Before risking any money at all in the stock market you must make sure that the financial necessities are taken care of. I'm not suggesting that you pay off the mortgage first, but do make sure that you have a pension policy, all the insurance you need and enough rainy-day money in a bank or building society account to cover any emergencies you might face. After that you can start with as little as you want. The cost of buying shares is weighted against very small deals so, as a rule of thumb, if you don't have over £1,000, and preferably nearer £3,000, to play with, try a different game.

Finding a stockbroker

Buying shares is a bit like buying apples. You wouldn't go to a wholesaler every time you felt like a little fruit. In the same way, you need a middleman, a stallholder if you like, to deal in your shares for you. And that man, or woman, is a stockbroker.

So your first step is to find yourself a broker. Nowadays they tend not to come in pinstriped suits and bowler hats – that stuffy image is gone for ever. They will be competing against each other for your business so don't be frightened of them.

If you're going to be starting out with less than £10,000 the field narrows down a bit. The very large firms of London stockbrokers won't want your business – they deal almost exclusively with the very large private clients and the institutions – so don't give it to them.

Ask around among your friends to see if any of them already has a broker they're satisfied with, and write to The Stock Exchange, Old Broad Street, London EC2N 1HP (tel. 071 797 1000) for a copy of their 'Private Investors' Directory', a list of brokers who specialize in dealing with private clients. You can tell by reading it which are most likely to be happy with your size of deal. The brokers offering 'international investment expertise' and 'personalized portfolio management' are less likely to help you with small bargains than those offering minimum commissions of £15 to £20.

Don't necessarily choose a broker based in London or near to where you live. Most of the dealing you do will be by phone so, barring the cost of the phone bills, it doesn't really matter where your broker is based. All deals in the stock market are done on the 'my word is my bond' rule. That means that once you ask him to buy or sell for you, you're committed. And if you don't pay up, the stockbroker will have to do so for you. So when you first get in touch with a stockbroker, by letter or phone, he will want to know a bit about you, financially.

In the first instance, he'll probably want a bank reference to check that you are financially solvent. But once he has made that check and is satisfied that you are sound, all you need to do is ring him with your instructions and he'll do the deal immediately.

Many brokers, particularly those outside London (known as country brokers), are also happy to chat to you about specific shares and offer advice on what to buy and sell. And they will offer the cheapest deals if you are buying small amounts of shares. Minimum commissions can be as low as £15–£18, and £20–£25 is the average minimum.

But time is money, and the more you have to invest, the longer he will spend advising you.

Other brokers specialize in a no-frills service. They will do the deals, but won't spend time chatting to you about it. And in exchange they keep their charges down. The Stock Exchange will be able to advise you which brokers offer this service.

In general you can choose from three main types of services:

1 *Execution only*. The broker will follow your instructions to buy or sell, but won't offer advice. Charges will be lower.
2 *Advisory*. You make the decisions following – or ignoring – the advice of your broker.
3 *Discretionary*. The broker can deal and then tell you what he has done. He would of course be dealing within parameters about the long-term objectives that you want from your portfolio but by being able to do so without telling you first he can work quickly and take decisions in an instant that may make (or lose!) you a lot of money.

Banks and building societies

Most banks and several building societies will be happy to do your deals for you. A few years ago I wouldn't have recommended it. Now I think you might get some of the best deals available by just walking down your local High Street.

Barclayshare, the computerized share-dealing arm of Barclays Bank, was one of the first in the field with a fast, efficient and cheap service, while National Westminster with its NatWest Stockbrokers pioneered the instant dealing on touch screens in its branches.

What these systems do is bypass the historic drawback to dealing in bank and building society branches – having to make contact through an employee who is not a specialist in buying and selling shares. Nowadays it is all done centrally and is linked to the branches through computers.

Many banks and building societies now offer competitive minimum and maximum commission charges which greatly undercut the large London stockbrokers – but check in advance just in case there are additional costs, or a high minimum commission.

HOW TO BUY SHARES AND SELL THEM AGAIN

Buying and selling shares is a simple process once you have got a relationship with a broker. For most people it is simply a matter of using the phone.

Buying shares

- Decide what you want to buy and how much you are prepared to pay for your shares.
- Ring your broker and ask him the current price of the shares. He'll be able to give you an instant answer by looking at his TV screen, which has all the share prices on it.
- If you want to go ahead, place your order.

At this stage you are committed to buying the shares – you cannot go back on your word. Your broker will then buy your shares for you at the keenest price possible, though it may be slightly different from the price he originally quoted to you. If you want him to ring you back before he buys them to tell you exactly what they will cost, ask him to do so; otherwise the deal will go through on your telephone instructions.

You can also ask him to report back to you on how much he actually paid.

The next day you will be posted a contract note confirming the details – check it to see that all's well. And a few weeks after that the company whose shares you have bought will send you a share certificate. This is your proof of purchase, so keep it safe.

If you lose your share certificate you will find that it is an expensive and bothersome business to get it replaced. Initially, write to the registrar of the company and he will take you through the steps. Far and away the easiest thing is not to lose it in the first place. Keep your certificates in a safe place at home, or you can pay for a bank or lawyer to keep them for you.

At best

If you think the shares in the company you fancy are going to be a real winner and you're not too fussy about how much you have to pay for them, ask your broker to buy them 'at best'. He will look at his screen, which has all the current prices on it, and get the best price for you, though it might be a few pence higher or lower than what you expected to pay.

New issues

Buying shares in new companies, either government sell-offs or companies coming to market for the first time, is easier still. You'll need a prospectus, which your broker will send you, or you can watch out for the main details, and the application form, being published in the quality papers.

All you do is fill in the simple application form, attach a cheque for the correct amount of money and send it off. Do it carefully. If you make any mistakes – often even minor ones – you won't get the shares.

If the offer is oversubscribed – that means investors have applied for more shares than are on offer – your application will either be scaled down, or put into a lucky dip known as a ballot.

Selling shares

- Decide which of the shares you own you want to sell, and make sure you know exactly how many you want to deal in. You don't have to sell all the shares you own in a company at the same time.

- Ring your broker and ask him the current price of the shares. Tell him how many you have and that you are a seller.
- If you want to go ahead place your order.
- Send your broker your share certificate and sign and return the transfer form which he sends you.

Paying the bills

Unless you're applying for a new issue, you don't have to pay for your shares immediately. All bills are settled on what the Stock Exchange calls Settlement Day (or Account Day).

The Stock Exchange year is divided into fortnightly account periods (occasionally they run to three weeks over Christmas and Easter and some bank holidays). And all the deals, whether buying or selling, in that account are settled on the same Settlement Day, usually six working days after the end of the account.

If you're a buyer you can have up to three weeks to find the money if you deal right at the start of an account, and if you're a seller you could have to wait that long for your money.

If you buy and sell within an account, you don't have to come up with the money at all. You just get a profit or stand a loss on Settlement Day. And your commission charges may be lower.

The cost of buying shares

Apart from the actual price of the shares, there are two main charges levied on you when you buy or sell:

- the dealer's turn
- the stockbroker's commission.

The dealer's turn. When you ask the price of a share, your broker will quote you two prices: 158–160p (and all shares are quoted in pence). That means that if you are a buyer you'll have to pay 160p, and if you are a seller you'll only get 158p per share. The difference of 2p is the dealer's turn, or the spread – the profit he makes out of dealing in the shares.

The price of the share quoted in the papers would then be the middle price of 159p. So when you're checking share prices remember to take into account the spread – and it can be a lot more than 2p.

The stockbroker's commission. Your broker will charge you a fee for his services, known as his commission, and it is added on to all deals, whether buying or selling. Although commission rates are no longer laid down by the Stock Exchange, most brokers charge between 1½ and 2½ per cent of the value of the deal, with a minimum for small deals.

This minimum can vary from around £15 all the way up to £50 or so if the broker wants to discourage small deals.

You'll also have to pay stamp duty at ½ a per cent and sometimes a bargain charge (see p. 29).

Example

The cost of buying 200 shares at £10.50 a share:

200 × £10.50	£2,100.00
Commission at 1.95%	40.95
Stamp duty at 0.5%	10.50
Bargain charge	3.00
Total bill	£2,154.45
Total costs	£54.45

MAKING MONEY

With luck and judgement the shares you buy will make you money. They can do this in two ways.

They might pay you a dividend twice a year. A dividend is a share of the company's profits, and the better the company is doing, the higher the dividend should go. The level of the dividend, as a percentage of the share price, is what is known as the yield on the share. And the yield provides income.

But the main way to make money on your shares is for them to rise in value – and this gives you capital growth. The higher the share goes, the more you make. But remember, a profit is only a profit when you sell your shares. So don't rely on paper gains.

Deciding when to sell

It is always worth making yourself a rule of thumb before you start out investing in shares so that you know when to sell. Decide how much will be enough of a profit for you, and stick to that rule. It might be when the shares rise 50 per cent, or double in value, or treble, or it might be when they have gone up and start to fall again.

And make yourself another rule on when to sell if the shares are falling in value. Don't hold indefinitely just in case the shares go up again.

Decide whether you'll stand a 10 per cent loss, or 20 per cent, maybe even 50 per cent, but once your shares reach the level you decided on, sell them.

Two old City maxims might help you here: 'Cut your losses but let your profits run' and 'It's no sin to take a profit.' It's up to you which one you use.

TAX

If your shares do make you money, either through a good yield or capital gain, you may have to pay tax on that profit:

- income tax on the dividends
- capital gains tax on any profit from selling shares.

Income tax

Income tax at 25 per cent is automatically deducted from all dividends before they are paid to you. But you must still declare them on your Inland Revenue tax return. If your top rate of tax is 40 per cent you'll be charged the extra. But if you are a non-taxpayer or pay tax at 20 per cent you can reclaim all or some of the tax that has been paid. See Chapter 10, Tax and Your Investments.

Capital gains tax

Capital gains tax is payable on all profits made on the sale of shares. However, the first tranche of profit is CGT-free. The amount of profit that is CGT-free changes in the Budget most years and the level is currently £5,800. So if your profit in the current year is less than £5,800 you'll get it tax-free. If it is more, you will have to pay tax on the difference at your top rate of tax, either 25 or 40 per cent.

But if any of your shares are sold at a loss, you can offset this loss against your profits, to work out how much you are due tax on.

So it is always worthwhile spending a couple of hours in March (before the end of the tax year on 5 April) working out how much profit you have taken in the year. If it is marginally more than £5,800 you may be able to sell a share that is showing a loss to bring you back down below the £5,800 level. Or you can do what is known as bed and breakfasting. You sell the shares one afternoon before the close of business and buy them back the next morning. That allows you to make use of the CGT allowance, either by taking a tax-free profit of up to £5,800 or showing a loss on a share to offset against your other profits, but still holding on to the share if you think it will come back up again.

A husband and wife now each have their own capital gains tax-free allowance.

Your profits from shares can also be 'indexed'. This means that the Inland Revenue allow you to include inflation in your sums before paying tax. It is a complicated concept but I explain it fully in Chapter 10, Tax and Your Investments.

WHAT TO BUY

This is the trickiest part of all – deciding which of the 6,000 or so shares traded on the Stock Exchange are going to go up in value. You can either take the advice of experts, or follow your own nose.

Expert advice

Your broker will keep you in touch with shares he thinks are going to go up. If you have a country broker he'll probably be particularly aware of what local companies are doing and may spot a potential winner before everyone else. But you won't be his only client – he'll be telling a lot of others as well, so the price of the share may go up before you buy it.

The daily and weekend papers are also full of share tips and astute reading of them may point you towards a few good buys. But remember these tips come, in the main, from the City dealers, who may have a good reason for tipping a share, and their reason may be to get the share price up so that they can sell. You may end up buying in a false market.

Tip sheets are another option. They make their money out of subscriptions, which can be costly, but because they have to do well in order to keep their subscribers their record tends to be fairly good. But here you have to make enough to cover the cost of your subscription, so unless your investment stake is quite high give tip sheets a miss.

And one other thought to keep at the front of your mind: if all the experts were as good as they think they are, they'd be sunning themselves on a beach in Barbados on their profits, not trying to sell you a tip.

Follow your nose

The chances are that you might be better placed than any of the experts to spot a potential winner.

Start by narrowing the field. Look at companies in the areas you work in and you might find that you can spot the ones that are picking up all the new business and likely to do well. After all, a company's share price will go up if its profits go up.

When you're out shopping – are any of the stores doing particularly well, always busy, suddenly cashing in on a new line of products? If so that might help their share price too.

On your way back from holiday – was the plane full of people saying 'What a good operator, I'll definitely book with them next year'?

Keeping your eyes and ears open can give you an insight into next year's high flyers.

KEEPING TRACK OF YOUR SHARES

Following the fortunes of your shares once you have bought them is relatively easy.

If you want hour-by-hour prices, you can always ring your broker – though he will stop taking your calls after a few days.

Some daily television and radio programmes also give share prices: Prestel, Ceefax and Oracle run a share-price service for subscribers, and British Telecom has a recorded prices and stock market bulletin on 0898 121220, updated hourly, and a more specific share price service, Teleshare, on 0898 500500. You pay premium rates for all these calls.

For most people a day-by-day guide to prices is enough. And you'll find that in the papers.

The better the quality of paper you buy, the more prices you'll find. And unless you've bought something really obscure you'll definitely find your share price quoted in the 'pink 'un', the *Financial Times*.

You'll actually get a lot more information than just the share price, but once you get the hang of interpreting all the figures you'll be able to monitor the performance of your shares automatically.

The FT index

It is always worth comparing how your shares are doing with how the market in general is performing. You don't have to monitor every single share or even the shares of the largest and most representative companies – that's all done for you. And the result is the *Financial Times* indices. Although the indices are named after the *Financial Times* newspaper, they can be found elsewhere. All the newspapers, most of the news bulletins – even the ones in the Jimmy Young programme – carry the latest update. There are three major indices:

- the *FT* 30-Share Index, which is the average rise or fall of a representative thirty shares in the market
- the *FT* All-Share Index, which monitors around 650 shares and is the most representative of the indices
- the *FT*-Stock Exchange 100-Share Index, which is the most often quoted of all. It is known for short as the *FT*-SE 100 and colloquially as Footsie. Even if your shares aren't in Footsie it's worth keeping an eye on it. If Footsie is going up and your shares are falling, against the trend, it might alert you that something is wrong with the company.

Annual report

Once a year you'll also be sent the annual report by the company you have invested in. If that appears too daunting for you, try to read the chairman's statement on future prospects at least. And, if you can, attend the annual meeting. If there is anything going wrong with the company, another shareholder may have spotted it and ask insistent questions of the directors. That too could give you the early warning you need to get out before everyone else does.

SPECIAL SITUATIONS

There is less jargon around in the City nowadays and for the small investor there are few technical terms you really need to understand. You can have a lot of fun and make quite a bit of money without ever needing to know about p/e ratios and arbitraging. But there are four special situations you are likely to come across and will have to understand:

Rights issue. This is when a company wants to raise extra money but doesn't want to increase its bank borrowings. It does it by offering shares to shareholders at a price cheaper than the market price. If the company looks sound it is a good time to buy on the cheap, but if you are not sure whether or not to buy, check the financial press on the day after the rights issue is announced. The comment there will tell you whether or not to go ahead.

A rights issue invariably depresses the market price of the shares, at least temporarily, because of the number of new shares on issue.

Scrip issue. This is a 'bonus' for shareholders which looks as if you get something for nothing. If a company has been doing well its share price will have been rising and once the shares reach a certain level, say 700 or 800p, the directors might think that that is too much. So they'll split the shares. With a 'one for one' scrip issue you'll double the number of shares you hold, but halve the price. So that instead of holding 200 shares at 800p, you'll end up with 400 shares at 400p. A scrip issue is usually very good for the shares and pushes the price up, though there is no inherent reason why this is so.

Takeover. If the company you have shares in looks attractive to another company it might attempt to take it over. It will make a bid, always offering more than the market price for the shares, and your directors will either advise you to accept or reject the offer. But it is up to you what you do. Your broker, or the papers, will offer plenty of advice, but the price of the shares usually rises the longer the bid goes on. It will fall, of course, if the bid fails, is withdrawn or is referred to the Monopolies Commission.

Suspended. The shares you own might be suspended at any time. This means that you cannot trade in them and the suspension can last for anything from a few hours to months. The reason is to give the company time to tie up a deal that has become common knowledge. If the deal is good for the company the shares will come back to market at a higher price, if it's not then they'll come back at a lower one. Either way there's nothing you can do but sit it out.

SHAREHOLDERS' PERKS

And here's a real bonus for shareholders. Many companies offer a sample of their products to shareholders, either free of charge or at a reduced price, to allow them to try out the goods or services.

And that means anything from cut-price Channel, or Chunnel, crossings to cheap dry cleaning. Hotel and restaurant chains often offer a discount on meals and overnights while many of the big High Street chain stores give shoppers a special discount.

For full details on discounts you can write to brokers Seymour, Pierce, Butterfield, 24 Chiswell Street, London EC1Y 4TY (tel. 071 814 8700) for a booklet on the subject (cost £5).

INVESTOR PROTECTION

Over the past couple of years, the government has put legislation in place that gives investors a high degree of protection against being taken for a ride by their adviser.

Of course no legislation at all can take the risk out of buying and selling shares. If the shares you buy go down, you lose money.

But they are tightening the rules to cut out any cowboys or conmen who set out to part you unscrupulously from your cash. For full details on investor protection see Chapter 30, Professional Money Advisers.

You can use the Securities and Futures Authority's arbitration scheme. Start by writing to the Complaints Bureau of the SFA at The Stock Exchange Building, London EC2N 1EQ (tel. 071 256 9000). That bureau will try to conciliate with the firm of stockbrokers but if they fail they will pass the matter on to the Consumers Arbitration Scheme, providing the sum involved is below £25,000.

5 PEPs and Share Option Schemes

The ins and outs of personal equity plans and share option schemes · Who can buy · Who should buy

See also
- *Chapter 4, Buying and Selling Shares*
- *Chapter 10, Tax and Your Investments*

PERSONAL EQUITY PLANS

PEPs, as they are known for short, were invented by the then Chancellor, Nigel Lawson, in 1986. What is on offer is income tax and capital gains tax relief for people wanting to invest in shares.

The trouble was that originally Lawson didn't give PEPs room to breathe. You couldn't invest enough cash to make them a worthwhile proposition for the fund management company to run at a profit, so charges were high, and you had to be a fairly long-term holder to get the benefits. So too few investors took the plunge, and PEPs just started to fade away. In his 1989 Budget Lawson stopped the rot. PEPs came back with a bang.

The ground rules

A general PEP allows you to invest up to £6,000 a year (£12,000 for married couples), and any or all of it can go into unit or investment trusts. At least half of your money has to be invested in European equities by the fund manager – either directly or through unit or investment trust holdings in European equities – but some can be in cash. The rest of the red tape that previously hindered fund managers and upped charges has been swept away.

On top of the £6,000, you can invest a further £3,000 in what is known as a single company PEP. That means you put the whole lot into one company. This is often used by people investing in the company they work for.

Who should join

Personal Equity Plans don't suit everyone. The tax benefits are geared more to the high-rate taxpayer who already successfully plays the stock market and already pays capital gains tax. However, investors prepared to put some money aside each year for a PEP could in the long term build up a good tax-free lump sum.

Because all your money can go directly into special unit and investment trust PEPs, these are now much more tax-efficient than going directly into unit and investment trusts.

Mortgage PEPs

The increase in the amount of money you can put into a PEP has spawned a new type of mortgage, the PEP mortgage.

There are three reasons why it is better value than an endowment mortgage:

- it has more tax advantages
- it is cheaper to buy because usually you pay less commission
- you can cash part of it in at any time without incurring penalties.

Only two things will stop it from taking over completely from the endowment mortgage – it pays less commission to the person selling it, and PEPs will still have to be explained to most people for a long time to come, whereas endowments are now well known and understood.

If you put your entire allowance of £6,000 a year – that's £500 a month – into a PEP to fund the capital part of the mortgage, you should, over twenty-five years, be able to cover a loan of around £300,000. Of course, you would still have to pay the interest on the loan to the building society or bank that lent you the money as well.

Choosing a personal equity plan

Most building societies have linked up with specialists to offer PEPs, while the major banks all have their own schemes; stockbrokers, financial consultants, unit and investment trust groups and some merchant banks offer PEPs. So you choose the one that suits you best and either phone the company or write to it directly, or use a financial intermediary. Selecting a PEP is more difficult. No one knows how well fund managers are going to perform in the future, so it is difficult to say which will be the best for you. However, decide in advance how much you want to put into unit or investment trusts, and that should narrow the field down.

If you simply want to put your money into a unit or investment trust PEP, choose a well-known fund-management company with a good track record. If you prefer to have your cash invested directly in shares, go instead for a PEP offering this service. The more discretion you have in choosing the shares, the higher will be the management charges.

SHARE OPTION SCHEMES

Share option schemes, as an investment, work on the principle of heads you win, tails you don't lose.

So, if you have the opportunity to join a scheme, you should take it. Not everyone can – you have to work for a company whose shares are quoted on the Stock Exchange, or a company that is a subsidiary of one that is quoted.

The schemes are designed to make you more loyal to your company. The theory is that by offering you shares in the company you work for – you will get the shares five years after you join the scheme – you work harder for that company now. The value of shares goes up with the level of profits so the better the company does, the higher the share price rises.

Buying your company shares through a share option scheme rather than just investing through the stock market is to your advantage for three reasons:

- there are good tax advantages
- you nearly always get the shares at a discount
- you don't have to pay for five or seven years, so you don't have to put any money up front.

There are two types of share option schemes:

- Save As You Earn
- executive share option scheme.

Save As You Earn

The most popular type of scheme is the Save As You Earn share option scheme.

Your company gives you the right to buy its shares at the current market price, or more often at a discount of up to 20 per cent. You don't get the shares there and then, they are held in trust for you while you save up the cash by making regular monthly payments into an SAYE building society or National Savings account. The company will tell you which building society to use.

You can save between £10 and £150 a month for either five years or seven years.

At the end of the given time you get a tax-free bonus added to your lump sum. If you have saved for five years you will get a bonus equal to twelve monthly payments. If you had decided on seven years, you leave your savings in the account, and at the end of the two years, you will get another bonus of a further twelve monthly instalments.

And it is at this point that you become the real winner. You choose what you want to do with the money. Obviously if the shares in your company have gone up in value in the five or seven years you'd be a fool not to take up the option to buy them at the original price and make a nice fat profit.

If the shares have gone down in value, you can choose not to take up your option. You just keep your savings, and the tax-free bonus.

Tax-wise you are on to a winner. No income tax is payable at all on the bonus you get. All you will be liable for is capital gains tax on the profits from the sale of your shares. For most people the sums involved are not likely to take them into the CGT bracket so all the gains will be tax-free.

Once you've been given the shares they are yours to do what you want with, you can sell immediately or hold as long as you like. You will receive dividends and can vote at your company's annual general meeting.

To qualify for a share option scheme you need to have been with your company for a certain number of years (usually three or five years) and work full time.

Redundancy, job-changing

If you leave work voluntarily, you will usually lose your right to exercise the options on your shares. You won't of course lose the money in the savings scheme and its interest.

If you've started out on an SAYE share option scheme and you lose your job through redundancy, injury or disability, or you reach retirement age and give up work, you have six months to take up your option to buy shares. You can buy as many shares as your SAYE scheme and its interest allows, but you cannot add any other money to it in order to buy more shares.

The rules are not the same in all company schemes, so if you are leaving work for whatever reason, check the arrangements in your scheme.

Takeovers, liquidations

If your company has the misfortune to go into liquidation during the time you are saving to buy shares, there would be little point in your exercising your option to invest in it. Under these circumstances you just withdraw your savings plus interest or their bonus. This money will of course be paid tax-free.

If the company is taken over the rules are more complicated and there can be an income tax liability. So you'd be best advised to check with your company to find out exactly where you stand.

For further information read Inland Revenue leaflet IR38, 'Income Tax, SAYE Share Options'.

Executive share option schemes

There is another, more up-market type of share option scheme which doesn't have the Save As You Earn savings plan attached – the executive share option scheme.

It works on the same principle of being offered shares at a price now that you have the option to take up in the future – that is between three and ten years of the option being granted. This share option scheme is aimed more at directors and senior managers, though other employees can be invited to join. The company can grant you shares worth up to £100,000 or four times your annual salary though, in practice, most companies will offer options worth much less than that.

Once you have exercised your option and the shares are yours, you can hold them or sell them, as you like.

To obtain the maximum tax advantages, you can exercise your options only once every three years.

Any profit you make between the price of the option and the price you sell the shares at is liable to capital gains tax if you make enough profit to come into the CGT net in any one tax year.

6 Life Insurance

The types of life insurance available · Which is the best one for you · How to buy it

See also
- *Chapter 14, Mortgages*

Life insurance often comes well down your list of financial priorities. By the time you have paid for insurance for the house, its contents, the car, your holiday, maybe even your bank loan and credit card bills, there may not be a lot left over for insuring yourself.

But if you have dependants – either a family of your own or elderly relatives – who rely on you to pay their bills, then you should have some sort of life insurance.

Life insurance works like this. You pay a set premium every month or year to a life company, which invests your money. If you die at any time during the term of the contract, the life insurance company pays out a lump sum to your estate; if you don't die, the insurance company pays you a large lump sum at the end of the contract. There are almost 300 life companies in this country and all of them are covered by the Policyholders Protection Act, which guarantees up to 90 per cent of your lump sum.

Nowadays the policies they offer are very sophisticated and come in all shapes and sizes. They can be tailored to fit most people's needs. But there are really only three basic types:

- term
- whole of life
- endowment.

And there's one other type worth adding in here:

- annuity.

TERM

This is the cheapest and most basic form of life insurance. You pay a small monthly or annual premium for a certain number of years – and it is up to you how long you want it to last – and if you die a lump sum is paid to your estate. If you don't die, you generally receive no refund of premiums or lump sum payment. This is what makes it different from other types of life insurance – and what makes it cheaper.

The longer the period you want to cover, the higher the lump sum you want to have paid out and the older you are when you start the policy, the higher the premiums.

Example

A healthy man wanting to cover a £20,000 lump sum for twenty-five years would pay:

£4.00 a month if he was	20 when he started the policy
£6.20	30
£11.80	40

All life insurance cover becomes more expensive as you get older for the simple reason that you are more likely to die and so cause the insurance company to pay out the lump sum. To cover that increased risk, the premiums are set at a higher level.

Aids

Aids is the spectre that haunts life insurance companies. Worried about the increasing number of deaths, and the forecasts for the spread of this incurable disease, most life insurance companies are now dramatically increasing the cost of term insurance to men. If you want to pay less, you will have to accept that the policy won't pay out if you die from an Aids-related illness. Not all insurance companies will write this sort of policy.

Because women are less likely to contract Aids, their premiums are not so expensive. A healthy woman wanting to cover a £20,000 lump sum for twenty-five years would pay:

£4.00* a month if she was	20 when she started the policy
£4.00	30
£7.40	40

*Minimum premium applies here, and therefore benefits are based on minimum premium – the sum assured would be £23,000.

Decreasing-term insurance

A variation on term insurance, this type of policy is better known as a mortgage protection plan and is usually used to cover a repayment mortgage. It means that the lump sum that is insured reduces over the years to parallel the reduction in the mortgage as it is being paid off. It is slightly cheaper than a straight term insurance policy.

Example

To cover £20,000 over twenty-five years with a decreasing-term insurance policy would cost:

£3.50* a month for a	20-year-old man
£5.00	30
£8.00	40

*Minimum premium – sum assured would be £24,500.

Family-income benefit policy

This variation works just like a straight term insurance policy except that instead of paying a lump sum on death it pays a regular income to the end of the term.

Some term insurance policies can be extended at the end of their life. You can convert them, regardless of your state of health, and they are usually called convertible term and cost a little more.

Apart from providing cheap life cover, term insurance can also be used to provide funds for an expected inheritance tax bill.

If you have made a potentially taxable gift and estimate that your estate would incur an inheritance tax bill (see Chapter 11, Inheritance Tax) of, say, £20,000, you could take out a term insurance policy for that amount so that if you die within the term the policy would pay the inheritance tax bill and the whole of your estate would then go to your beneficiaries.

Tip

If you are using a term insurance policy, or indeed any sort of life insurance policy, to meet your inheritance tax bill make sure the lump sum passes directly to your heirs and doesn't become part of your estate or it will become liable for inheritance tax itself. This is quite easy to do – just use a standard form of trust. Full details of inheritance tax planning are given in Chapter 11.

WHOLE OF LIFE

Unlike other life insurance policies, whole of life always pays out. You pay premiums for the rest of your life and on your death a lump sum is paid.

The younger you are when you start the policy the cheaper it is. Most policies are with-profits policies.

Example

A low-cost whole of life policy to fund an expected pay-out of £20,000 would cost around:

£20.00 a month for a	20-year-old man for 30 years, then	£17
£20.00	30	£15
£20.00	40	£14

If you wanted to buy a with-profits whole of life policy so that your lump sum would be much greater than the £20,000 by the time it pays out, the monthly premiums would be heftier:

£33.00 a month for a	20-year-old man
£41.40	30
£53.40	40

The forty-year-old, on his death, could have an estimated £39,000 paid to his estate at age seventy-five if you assume annual growth of 10.5 per cent, £19,000 on annual growth of 7.0 per cent, with the low-cost whole of life. At eighty-five, he'd get £70,000 (10.5 per cent growth) or £27,000 (7.0 per cent). If he had been paying the higher premiums, the pay-out on death at age seventy-five would be £79,000 (7.0 per cent) or £169,000 (10.5 per cent), or £111,000 (7.0 per cent) or £290,000 (10.5 per cent) if he died at eighty-five.

Tip

Low-cost whole of life policies cost more than term insurance. But they are much better value because whole of life always pay out – after all, everyone dies; term policies only pay out if you die in the specified period.

There are variations. A limited payment policy suits people who don't think they will be able to afford the payments throughout their life. On this type of policy you stop paying premiums at a specified age – probably when you retire – but of course that means you will pay more during your working life.

The problem with both the term policy and the whole of life policy is that *you* will never get your hands on the money. You pay the premiums but it is your dependants who benefit from the pay-out unless you choose to surrender your whole of life policy.

ENDOWMENT

Endowment policies solve this problem because they pay out if you die – and they pay out if you don't. They are a sort of hybrid, combining term and whole of life insurance policies.

You decide on the lump sum you want and you pay the premiums to cover it. If you die the money goes to your estate, but if you survive to the end of the contract, then the money comes to you. Of course, this sort of policy is much more expensive than a term or whole of life policy. Many people use this sort of policy as a savings scheme or to run alongside their mortgage.

Because it is the most popular of the life insurance policies, the endowment has most variations.

With-profits

This policy guarantees the lump sum, which will be paid out if you die (that is known technically as the sum insured), and offers you an additional bonus when you reach the end of the term. Not only will you get the agreed lump sum but you should get a nice fat extra payment on top. Of course this all has to be paid for, and your premiums will be higher.

Low-cost

To cut the cost of your premiums you can go for this type of endowment policy. It guarantees a smaller lump sum and doesn't offer as much of a bonus. The advantage is that the premiums are much lower than the full with-profits endowments. This type of policy is the most popular one for use with endowment mortgages.

Low-start, low-cost

Premiums here are lower still. You pay less in premiums for the first five years of the policy – though they rise by 20 per cent every year – and stay at the higher level for the remaining years to balance out this reduction. Otherwise the policy works in the same way as the low-cost policy.

Example

Endowment premiums for a £20,000 lump sum, taken out by a man of thirty:

With-profits endowment	£70.60 a month
Low-cost	£27.00 a month
Low-cost, low-start	£15.30 a month (increase by 20 per cent every year for five years)

The estimated lump sum pay-outs from these premiums are:

	7%	10.5%
With-profits endowment	£47,000	£78,000
Low-cost	£17,000	£29,000
Low-cost, low-start	£17,000	£29,000

The bonus on the with-profits contracts is paid in two ways. There is an annual 'reversionary' bonus. Once it is added to your policy it cannot be taken away again so you know that your endowment has grown by that amount. And there is the 'terminal' bonus, which is added on when the policy matures. You won't know how much this is going to be until the very end of the contract so don't expect too much.

If you are using your life insurance as a form of saving there is another alternative – *unit-linked life insurance*. These policies give you the advantages of investing in the stock market or unit trusts – or whichever type of link you choose – with the additional benefit of life assurance cover.

You could get the same benefits of course by opting for term insurance and paying into, say, a monthly unit trust policy.

ANNUITY

An annuity is an endowment policy in reverse. It tends to be used by elderly people wanting to guarantee an income for the rest of their life, or for buying a pension.

They pay the insurance company a lump sum, and the insurance company guarantees to pay them a monthly or yearly income until they die. So the older they are when they take out the policy the more they will get every month.

Example

An annuity based on someone paying a lump sum of £20,000 would fund an annual net payment of:

£3,130 a year for a	75-year-old man
£3,500	80-year-old man
£3,950	85-year-old man

Couples can take out a joint annuity which will pay a regular income until the second death but it will pay less per month than a single life annuity would.

An annuity based on a couple paying £20,000 and wanting the payments to continue until the death of the survivor would pay out:

£2,200	if the man is	75 and the woman	72 when they buy the policy
£2,500		80	77
£3,000		85	82

The drawback with annuities is that if the holder dies immediately after taking out the policy, the lump sum is lost, and the payments stop on death. To cover this eventuality you can take out a different type of annuity which will continue to pay out to your estate for a guaranteed period, usually between five and ten years even if you die. Of course, this will reduce the income from the policy.

The seventy-five-year-old man paying a lump sum of £20,000 and guaranteeing the pay-outs for at least five years would get only £2,900 a year, instead of £3,150. If a husband was eighty-five and his wife eighty-two, the couple would receive £2,900 a year instead of £3,000.

Annuities are often used with home income and reversion plans so that the elderly person or couple can unlock the capital in their home to give them an income, while still continuing to live in the house. See Chapter 28, Housing for the Elderly.

There are variations on the standard annuity too:

Deferred annuities

These are bought now, but don't start paying out a regular income until later.

Capital-protected annuities

These give a lifetime income and some form of capital repayment on early death.

Temporary annuities

These give you a fixed income for however long you want and are often linked to endowment policies.

POINTS TO WATCH

Tax

Since March 1984 the premiums you pay to the life insurance company on new policies no longer qualify for limited tax relief. The premiums have to be paid from your taxed income.

The one exception to this rule is the life insurance that is tied into a personal pension plan. Premiums funding that policy are tax deductible.

Basic rate taxpayers don't pay income tax on the proceeds of any life insurance policy.

Higher rate taxpayers only get the proceeds of the life insurance policy – that is the money you get at the end – tax-free if it is what is known as a 'qualifying' one.

To be a 'qualifying' policy it has to:

- pay out at death

or

- run for at least ten years; if the policy is only designed to run for ten years then you must be at least seven and a half years into it
- you must also pay the premiums regularly – annually or monthly.

The company should tell you before you buy the policy whether or not it is a qualifying one.

> **Tip**
> If you have a life insurance policy taken out before March 1984 you will get tax relief on the premiums so don't cash it in and take out another one if, for example, you are moving house and increasing your mortgage. Stick with the one you've got and take out another, to make up the difference.

Front-end loading

The great disadvantage of life insurance is the size of the commission that is paid to the salesman. Up to a full year's premiums on a long-term policy can go to him, rather than into your policy. That means that if you cash in your policy early, you will get very little back. Certainly you won't even get your premiums back if you surrender your policy in the early years.

So never overcommit yourself with life insurance. If you don't think you will be able to keep up the payments, don't take out the policy.

A thirty-year-old man paying £20 a month on a twenty-five-year with-profits policy would get about:

7%	10.5%		
£355	£355	if he cashed in at the end of	2 years
£565	£565		3
£1,100	£1,100		5
£2,700	£7,500		10
£2,900	£11,500		20
£13,000	£21,500	if he allowed the policy to mature	

If you find you really can't afford the premiums and have to stop paying, opt for the 'paid up' alternative. That means you stop paying the premiums but the policy continues to run to maturity and then your money is paid out. The money that is in the policy should continue to grow during that time though the sum insured will be reduced.

Medical

All types of life insurance depend on the insurance company getting its sums right. That is the job of the company actuary. He uses statistics and tables of average life-spans to work out how long he thinks you will live – and that way sets your premiums.

Some companies will ask you to take a medical, others won't. It depends on your age and state of health and the amount of cover requested. If you are in perfect health it may be worthwhile going for a company that wants you to have a medical examination – they may be paying bigger bonuses than one which is not so fussy. But if you don't have to go for a formal medical examination, you will have to put all of your medical history down on the application form, and they will write to your own doctor in any case.

The bogey in the life insurance world at the moment is Aids. The worry is that anyone who contracts the disease will take out a life insurance policy to provide for their dependants. Since there is no cure yet for Aids, the company would have to pay up.

All policies now ask you direct questions about the disease. Don't lie just to get the policy. If you die, and the company finds out that an Aids-associated illness was the killer, they won't pay out the lump sum and your premiums will have been wasted.

Warning. Never withhold information about your health or activities from an insurance company. No say, no pay.

Tip

Non-smokers can get cheaper cover. So if you don't smoke cigarettes (cigars and pipes are not seen as killers) shop around for a life company which offers reductions. If you do smoke, ask if the company will reduce your premiums if you give up and sign a non-smoker declaration after twelve months.

Monthly term insurance premiums for £20,000 lump sum

Age	*Smoker*	*Non-smoker*
20	£4.60	£4.00
30	£7.40	£6.20
40	£15.00	£11.80

Women

All the examples I have used above are for men taking out life insurance. I have done this deliberately because in the main it is men – as the traditional breadwinners – who buy life cover. Because women are expected to live on average four years longer than men, life cover for them is cheaper – though annuities of course are more expensive.

BUYING LIFE INSURANCE

It is impossible to find out which will be the best company to buy life insurance from. No one knows what is going to happen in the future so you can't tell which will be the best performing fund. All you can do is opt for the life companies with the best track record and hope that their investment expertise continues to flourish.

There are three main ways of buying life insurance:

- through a tied agent
- from an independent financial intermediary
- from banks and building societies.

Tied agents

A number of the large companies which sell life insurance among their financial products, such as the Prudential, Abbey Life and Allied Dunbar, use company salesmen to sell their products. These men and women are tied agents of the company and can only sell that company's products. They will have to tell you at the outset that they are tied to one particular company. See Chapter 30, Professional Money Advisers.

In the past, salesmen (whether tied or independent) received their commission first, before any of your premiums went into the life insurance fund. That meant that if you had to surrender your policy in the first year or two you got very little back at all. But the rules have changed. Commission is now spread over more years, so early surrender doesn't cost the policyholder quite so dearly.

Independent financial intermediary

If a salesman is independent (that means he is not a tied agent) he has to offer you the best possible advice that he can. Legally he is not allowed to sell you a product just because it pays him a higher commission than any other.

He won't automatically offer to tell you how much commission he is getting but if you ask – and you should – he will tell you. You should also check that he is a member of the independent financial intermediaries association, FIMBRA, or the Insurance Brokers Registration Council, because that gives you quite a degree of investor protection. FIMBRA (Financial Intermediaries, Managers and Brokers Regulatory Association) members are covered by a compensation scheme which pays back 100 per cent of your money up to £30,000 and 90 per cent of the next £20,000 should anything go wrong. That means you get £48,000 of your first £50,000 invested if your adviser turns out to be a fraudster.

Check that he is a member of FIMBRA before you sign over any money. To do that phone FIMBRA on 071 538 8860. You'll need to know your adviser's name and his postal code. Or look up the FIMBRA list, which is available in most public libraries.

If the salesman is an accountant or solicitor he should also be a member of the Recognized Professional Bodies, which shows that he is qualified to give independent advice on life insurance.

Banks and building societies

If you are buying a life insurance policy through your bank or building society, perhaps, though not necessarily, as an endowment mortgage, you will find that it is either selling only its own products, or it is offering the full range of products on the market. It is wise to know at the outset which category your lender comes into.

Many of the larger building societies have now linked up with one insurance company or life office and offer only the products of that company. It means that if you are looking for, say, an endowment mortgage, you may not get the deal best suited to your circumstances. You would be better to choose the insurance company first. Go for one with a good record, and your bank or building society should accept it.

If you want to go to the building society first but don't want the agent to which it is tied, ask for the help of the financial advice subsidiary. All the building societies have one, or are in the process of setting one up, and if you use this subsidiary it will shop around for the best deal for you.

Once you have signed a life insurance contract you are not immediately bound to continue it. You have a cooling-off period of fourteen days after you receive the official letter from the life company. If you want to cancel, do it then.

There are two main exceptions to this rule. You get no cooling-off period at all if you are buying:

- some kinds of single premium policies
- some types of pension.

Your adviser (whether independent or tied) will tell you if you have a cooling-off period. So think before you sign.

COMPLAINTS

If you are dissatisfied with the service you are getting from your life company, or if you have a complaint about the policy, begin by complaining to the company itself. Start with the person who runs the branch office, and complain all the way up to the chief executive or managing director. If you still think that you are being unfairly treated there are two other courses of action open to you:

- The Association of British Insurers – the trade body of the insurance companies – can help you. They won't be able to arbitrate on your behalf but they will give you all the help they can. Write to them at 51 Gresham Street, London EC2V 7HQ (tel. 071 600 3333).

- The Insurance Ombudsman may be able to help you. If the company you are dealing with is a member of his bureau he will be able to sort out your complaint. His findings are binding on the insurance company but not on you. He won't be able to help you unless you have tried to have your complaint sorted out by the company itself. If that has failed you and you take a genuine grievance to him he should be able to do something about it for you. The Insurance Ombudsman can be found at 135 Park Street, London SE1 9EA (tel. 071 928 7600).

If your life insurance company is not a member of the ABI or the Insurance Ombudsman's Bureau there are other courses of action open to you – see pp. 197–8.

PART TWO

TAX

7 Understanding Income Tax

What are tax-free allowances? · The PAYE tax system · NI contributions · Breaking the tax code · How to fill in your tax return · How to pay less tax

See also

- *Chapter 8, Tax and Your Job*
- *Chapter 9, Tax and the Self-employed*
- *Chapter 10, Tax and Your Investments*

If you are looking for someone to blame, Napoleon is your man. For it was to finance the war against him that income tax was introduced, as a temporary measure, in 1799. It has been with us ever since. Income tax is a tax on your income and the money raised is used by the government to pay for schools, roads, hospitals, defence and the welfare services. If you have money coming in from your salary, pension, investments and even some DSS benefits you will be liable for tax on it.

You can't evade income tax but you can avoid paying any more than you have to by making sure you understand the system and taking advantage of all the allowances that you can.

In his 1988 Budget, Chancellor Nigel Lawson announced a major and radical reform of the way women are treated by the Inland Revenue.

For the first time in almost two centuries, women were given control over their own tax affairs: they have their own allowances and pay their own income tax on income. Married women are no longer penalized for stopping work to raise a family.

ALLOWANCES

You don't pay tax on *every* penny of income that you get. Everyone has a tax-free allowance of some sort. That means that the first tranche of your income is tax-free – how large that tranche is depends on your personal circumstances. A married couple are allowed to earn more before paying tax than two single people, and a married pensioner who is over seventy-five can qualify for the largest allowance of all.

The tax-free allowances are increased annually in the Budget, at least by the rate of inflation. At the moment that means that they go up by around £100 to £200 or so a year.

It is important to remember that an allowance is not a payment made to you by the Inland Revenue. It is the amount of money you can have as income before you start paying tax.

The main tax-free allowances are the personal allowance, the married couple's allowance and the age allowances. There are other allowances which can only be claimed in special circumstances.

Personal allowance = £3,445 in 1992–3

This allowance can be claimed by anyone resident in the UK who is single and has an income – it doesn't matter whether they are children or grown-ups, widowed, divorced, unmarried or separated.

The income can be from any source – your salary, pension, interest on your savings or dividends from investments. They will all be tax-free up to the level of the personal tax-free allowance.

Married couple's allowance = £1,720 in 1992–3

Since the Inland Revenue is not psychic you have to tell your tax inspector when you get married. You can then claim this allowance. In the year that he marries, the man gets a proportion of this allowance depending on which month he got married, and the months are rounded down in favour of the taxpayer.

This allowance is paid automatically to the husband, but after 6 April 1993, if the wife wants to claim half of it she can. All she has to do is write to her tax office and ask; she does not need her husband's approval or permission. But you do have to make your claim before the start of the tax year. If the wife wants more than her half – perhaps because she is a taxpayer and he is not, or a higher-rate taxpayer than her husband – then she does need his signature.

Until 6 April 1993, a wife will get the remainder of the allowance if the husband cannot use it all.

Age allowance

- Married couple's allowance aged 65–74 = £2,465
- Personal allowance aged 65–74 = £4,200
- Married couple's allowance aged over 75 = £2,505
- Personal allowance aged over 75 = £4,370

This is an allowance given to men and women aged over sixty-four. Women are completely equal on this one and don't get the age allowance when they reach sixty. They have to wait until they are sixty-four or over at the start of a tax year. The personal age allowance replaces the ordinary personal allowance and the married couple's age allowance replaces the ordinary married couple's allowance.

The theory behind the age allowances is that they help pensioners to manage on a limited income by giving them a larger slice of tax-free income. However, if the pensioners are well off then they don't need that extra slice, so the government takes it back again. The government's definition of 'well off' is having an income of more than £14,200 a year in the 1992–3 tax year. If the pensioner exceeds that then his age allowance is cut by £1 for every £2 of income over this level, until the allowance falls to that of the personal tax-free allowance.

The age allowance can be claimed when either partner of a married couple reaches sixty-five. That means they get the full allowance for the tax year in which their sixty-fifth birthday falls. So it is the year in which one of them is sixty-four or over on 6 April.

Once either of the couple, or a single person, reaches seventy-five the age allowance is stepped up. But again, when annual income reaches £14,200 they start to lose the extra allowance.

Current basic allowances 1992–3	
Personal allowance	£3,445
Married couple's allowance	£1,720
Personal age allowance	£4,200
Married couple's allowance	£2,465
Personal age (75 and over)	£4,370
Married couple's (75 and over)	£2,505

Additional personal = £1,720 in 1992–3

This allowance is equal to the difference between the married couple's and the personal tax-free allowance and can be claimed by anyone bringing up children on their own whether they are widowed, divorced, unmarried or separated. Married men whose wives are totally incapacitated can also claim this allowance. The child living with the single parent must be in full-time education or under the age of sixteen at the start of the tax year, or being trained by an employer for at least two years for a trade, profession or vocation. You can only claim the allowance once. However, if a divorced or separated couple each have one or more children living with them, both can still claim an additional personal allowance.

If the child or children live for part of the year with one parent, and part of the year with the other, the allowance can be split.

Blind person's = £1,080

If you are registered blind then you can claim this relief. If your spouse is also registered blind, you can claim double the relief. The blind person's allowance will be reduced by any disability benefit received but if you cannot use all the allowance, the remainder can be transferred to your spouse regardless of their sight.

Widow's bereavement = £1,720

In the tax year that her husband dies and the following tax year, a widow is entitled to this allowance, as well as the single person's allowance, providing she does not remarry in the meantime. There is no widower's bereavement allowance because, traditionally, women have been financially dependent on their husbands whereas husbands were seldom financially dependent on their wives. Whether the tax rules will be changed to allow for 'men's lib' remains to be seen.

OUTGOINGS

Knowing what allowances you are entitled to is only half the story. You can also get tax relief on certain outgoings, that is certain expenses you have in connection with your job (see p. 93).

Some of your outgoings automatically qualify for tax relief providing you tell the Inland Revenue about them:

- your pension contributions up to 15 per cent of your earnings if you are an employee or 17½ per cent if you are self-employed and under thirty-five. Self-employed people over thirty-five are allowed more. See Chapter 27, Pensions
- the interest on your mortgage payments on a home loan of up to £30,000
- investments in BES schemes (see p. 117), until December 1993
- money you spend on special clothing you need for your job if you provide it yourself, essential tools you have to buy, necessary reference books, travelling expenses incurred in doing your job (this does not include travelling to and from work), interest on loans to buy essential equipment and fees and subscriptions to professional bodies.

If you are not sure what you can claim try getting in touch with your union or professional association to find out what, if any, allowances they have agreed with the Inland Revenue.

Tip
Not all allowances have prior agreement so claim everything you think you are entitled to. If you are not entitled your tax inspector will delete what you shouldn't get. But if you don't claim your inspector won't insert anything.

PAYE

If you are an employee, your employer will deduct tax from your wages before you get them. The system is known as Pay As You Earn (PAYE). Any other income, whether the tax is deducted at source or not, will have to be declared to the Inland Revenue. You could have money coming to you from a secondary job, such as running a catalogue club or doing clothing alterations for friends, or you might have your savings in an account such as the National Savings investment account where the interest is paid gross, that is before tax is deducted.

So to work out how much tax you should be paying the Inland Revenue will start with:

Your annual salary	say £14,300	
Deduct your allowances	−£5,165	= £9,135
Deduct your outgoings	−£200	= £8,935
Add other income	+£2,400	= £11,335
And tax you on the result . . .	£11,335.	

This is known as your *taxable income*.

Not all taxable income is taxed at the same rate. If you have a taxable income of more than £23,700 in the 1992–3 tax year you will have to pay tax at 40 per cent on anything over that figure.

In the past there have been up to around half a dozen tax bands, rising from the basic rate in steps and stairs. Nowadays, it is down to three – 20 per cent on the first £2,000 of taxable income; 25 per cent on the next £21,700; and 40 per cent on the rest.

The tax bands are usually increased annually in the Budget, at least by the rate of inflation, often by more. If a Chancellor has scope to cut taxes he can do it in one of two ways:

- he can reduce the basic rate of tax, or any other rates
- he can raise the tax bands by more than the rate of inflation.

Or he can use a combination of both.

WORKING THE SYSTEM

If a husband works and a wife doesn't, then it is in their interest to have any savings in her name. A wife has a personal allowance which she can offset against any income – and that includes interest from a savings account. If her husband is paying tax – at whatever rate – the interest will be paid less tax to him. Or tax-free to her if she can offset it against her allowance.

So she will get more interest on the money because no tax is deducted.

But remember, if the cash is in a bank or building society it is crucial that the wife signs a form asking for the interest to be paid gross, otherwise it will be paid net and she will have to claim back the tax.

And if a wife is in a higher tax bracket than her husband (or pays tax where he doesn't), remember to opt for the married couple's tax-free allowance to be claimed by her.

NATIONAL INSURANCE

The other obligatory 'tax' you will have to pay, if you have a job, is National Insurance. Your contributions, and those of your employer, entitle you to state benefits such as pensions, unemployment benefit, maternity and redundancy pay.

How much you pay depends on:

- what you earn
- what your status is.

What you earn

If you earn between £54 a week (the lower earnings limit) and £405 a week (the upper earnings limit) you will pay National Insurance contributions as a percentage of your salary.

You don't pay National Insurance

- if you earn less than £54 a week
- on anything you earn over £405 a week.

And if your employer's pension scheme is contracted out of the state pension scheme – that is, your employer's scheme guarantees you a better deal than you would get from the earnings-related part of the state pension – you pay lower NI contributions.

Contributions are on a sliding scale and start once you earn over £54 a week:

Weekly earnings	*Contracted in*	*Contracted out*
Up to £54	2 per cent	0 per cent
£54–405	9 per cent	7 per cent

What your status is

There are four classes of NI contributions:

Class 1. For employees earning between £54 and £405 a week. Both the employee and the employer contribute (though there is no upper limit for employers) and providing contributions are up to date the employee qualifies to claim most state benefits.

Class 2. For the self-employed with annual profits of more than £3,030. The NI contribution is a flat rate of £5.35 a week and that entitles the payer to claim all benefits except unemployment and industrial injuries and will not go towards any SERPS entitlement on his pension (see Chapter 27, Pensions).

Class 3. Allows anyone with gaps in their NI record to make voluntary payments at a flat rate of £5.25 a week. By making up your record you may then qualify for benefits such as the full basic state pension, widow's benefits, the death grant and so on.

Class 4. For the self-employed. In addition to the flat rate Class 2 payments, the self-employed must pay a further 6.3 per cent of any profits they make between £6,120 and £21,060.

These figures change every year.

TAX CODE

The last thing anyone wants to do is pay too much tax. Or too little; if you underpay you'll get a large bill if the Inland Revenue catches up with you, so it is just as bad as paying too much.

If you are taxed under the PAYE system, make sure that you are paying the right amount of tax by checking your tax code.

Your notice of coding is sent out in January or February for the next tax year. Not everyone gets one every year, but you can find out what yours is by looking at your payslip. It is always shown there. If you think that the amount of tax you should be paying has changed from the previous year, perhaps because you got married, or you stopped your lucrative little sideline, or you have been given a company car, then make sure your tax code is changed.

Do this by writing to your tax office to tell them of the changes.

Cracking the tax code

Your tax code is made up of two parts:

- a series of (probably) three numbers
- a letter.

And it will look something like 344L.

The letter

That tells you and the tax man who and what you are:

L a single person or married woman
H a married man or single parent with a child in full-time education
P a single person under seventy-five getting the full age allowance
V a married man getting the full age allowance

T you don't fall into these categories, don't want your employer to know your status, or are a single person over seventy-five getting the full age allowance
K the Inland Revenue is collecting tax you owe from previous years
NT you pay no tax at all
OT all your pay is to be taxed because you have used up all your allowances.

The numbers

This shows how much you are entitled to in allowances. The tax man will start with the basic allowance such as the personal allowance and add on any others that you are due. He will also include here any allowances you are entitled to for, say, special clothing or tools you have to buy.

From that he will deduct any additional tax that you owe him, from say the interest on a National Savings investment account, tax on the perk of a company car or tax from a small sideline job you have.

And the figure he is left with he divides by ten, by knocking off the last digit and replacing it with the letter. That gives you your tax code.

To do it the other way round and get the monetary value of your tax code – add zero after the last number: 344L becomes £3,440.

Check the code

The notice of coding is sent to you and to your employer, who uses it to work out how much tax should be deducted from your salary. Check it, or you could end up paying the wrong amount and you could be faced with an extra bill at the end of the year. There is no point in telling your employer if it is wrong – he has to use the code the Inland Revenue gives him for you. So tell the Inland Revenue.

You should also tell the tax man if your status changes – if you retire, get married or get divorced. Any of these changes may affect the personal allowances you are entitled to and will therefore affect your tax code. It will need to be changed.

FILLING IN YOUR TAX RETURN

Don't expect a tax form every year. If your affairs are relatively straightforward you will only get a tax form every few years. But if your circumstances change, tell the tax man. Write to him outlining what has happened and he may send you a form, or just change your notice of coding. You are legally bound to tell him of any new source of income, such as earnings from other work that you do. If you don't, you could be fined.

There are several different types of tax form:

- Form P1: for people with simple tax affairs
- Form P11: for people with more complicated financial affairs
- Form 11: for the self-employed.

But all forms work on the same principle – they look back over the past financial year, and forward to the next one.

The tax year runs from 6 April of one year to 5 April of the next. So the 1992–3 tax form will be for the year 6 April 1992 to 5 April 1993 – that means it will look back on your income and capital gains for the year 1991–2 and look forward to your allowances for the year 1992–3.

Every tax form comes with guidance notes, which are worth reading before you fill in the form. The form and the notes have won a plain English award for being easy to understand, so there are no excuses for not filling in the right columns.

It really shouldn't take more than a wet Sunday afternoon to fill in the form.

The tax form

Now that everyone is seen as a separate entity by the tax man, we all get our own tax form to fill in – if we get one at all. That means that a wife will get a separate form rather than just a different column on her husband's form. If a wife doesn't want to fill hers in herself, there's no law that says she must. She can ask her husband to do it, if that is what she wants, or an accountant or anyone else for that matter. But she has to sign it, and it is her responsibility to ensure that all the information on it is correct.

The tax form is divided up into main parts, and you should go through each part with the attached notes to hand to make sure that you are doing it correctly.

1 Earnings

The figure you fill in here is your total earnings for the year, including overtime, bonuses, commission, holiday pay and so on, less your contributions to a pension scheme. You should get your earnings figure from your employer at the end of every tax year on a Form P60.

Tips are taxable. So if you get tips in your job as a hairdresser or taxi driver you will have to pay tax on them. Some unions have already agreed a figure with the Inland Revenue on what the likely annual tips will be for a particular job, such as milkmen. And it is that figure, rather than the actual amount of tips you made, that you will be taxed on.

If you get fringe benefits from your employer, such as a company car, or free membership of BUPA, you will have to pay tax on the value of the perks if you earn over £8,500 a year (see below). Your employer will send a note of your benefits to the Inland Revenue on Form P11D. You should fill in the value of the perks in this part of your form. Remember to include your company car here if you get one.

Tip
Check with your employer that your figures match the ones he is sending in. Ask him for a copy.

If you leave your job during the year either through enforced or voluntary redundancy or early retirement and you get a lump sum payment, it may be taxable (see p. 302). Fill in the figure in the appropriate box and the Inland Revenue will work out whether or not you are due to pay tax on it or, if it has already been taxed under PAYE, whether you are due a refund.

If your employer subsidizes your travel to and from work or pays you luncheon vouchers of more than 15p a day, you will have to give the Inland Revenue details.

You must also tell the Revenue about any company profit-sharing scheme.

If you received DSS benefit or unemployment benefit tick the appropriate box. You don't need to give the amount on Form P1 – though you will on Forms P11 and 11. Not all supplementary and unemployment benefit is taxable – though some of it is – so ask your benefit office how much of yours is subject to tax.

2 Pensions

This part of the tax form is for people receiving pensions or about to retire. If you are not in that category move on to the next part. If you do get pension give full details of everything you receive – not all pensions are taxable. In general supplementary pensions, invalidity and disablement pensions and the £10 Christmas bonus are tax-free, as are many of the pensions linked to war service and injury, but check with your tax office if you are not sure or you think you are being unfairly treated.

If you get the widow's pension, widow's payment or widowed mother's allowance, an invalid care allowance or industrial death benefit, give details here.

It is particularly important to be accurate here because it is a pensioner's nightmare to pay the wrong amount of tax. Either you are paying too much at a time in your life when you can least afford to, or you are paying too little and when the Inland Revenue realizes that and asks for the money it will wreck any budgeting you have tried to do.

3 Investment, savings, etc.

This part of the tax form is fairly self-evident. If you have savings and investments or income from any other source that you haven't already mentioned on the tax form you complete the appropriate box. Although interest from building society and bank accounts is now taxed at the basic rate before you get it, you still have to fill in how much you received.

It is important to the tax man if:

- you pay tax at more than the basic rate, because you will then have to pay more tax on the interest
- you are over sixty-five because if you breach the age allowance income barrier of £14,200 you will have to pay extra tax.

If you deal in shares, keep your dividend receipts and fill in how much you received from each dividend. If there is not enough room, put all the details on a separate piece of paper and attach it to your tax form.

If you had income from trusts, interest from a local authority loan or dividends from gilts, put the amounts you received in the box marked 'other dividends, interest, etc.'.

4 Outgoings

Outgoings are the payments you make which are tax deductible. If you pay your income tax through the PAYE system there are not many expenses which you have which you will be able to offset. However, if your expense was 'wholly, exclusively and necessarily in the performance of the duties of your employment', then it will be deductible. Some tax inspectors are very rigid about this rule and if you buy something, such as a word processor, which is necessary for your work, but which you also want for your own use, they won't allow you to offset the maintenance and running costs because it is not 'wholly, exclusively and necessarily for work'. (The actual cost of the computer would be a capital cost and therefore not offsettable anyway.)

Other tax inspectors may allow you to divide the running costs between work and home use and charge the work portion against your tax bill.

If your union has already agreed a 'fixed deduction' with the Revenue for special clothing or tools you have to buy, then you can claim this even if you spend less than the agreed amount. If you spend more, you can try to claim more.

Subscriptions and fees to professional bodies means what it says – union dues are seldom allowable.

5 Capital gains tax

If you buy and sell shares or sell a second home or an expensive antique dining-room table you could be liable for capital gains tax. However, the first tranche of profit is tax-free, as are any items you sell providing that, in total, they are not worth more than a certain figure. You will not be liable for tax in 1992–3, if your gains are less than £5,800, but any profit you make above that figure will be taxed at your top rate. See Chapter 10, Tax and Your Investments, for full CGT details.

If you have made capital gains and look liable for tax, it might be worthwhile getting the advice of an accountant or professional financial adviser.

6 Allowances

See above (p. 67) for who can claim which allowances. If you qualify, tick the appropriate box.

7 Declaration

And this is the easy bit. If you have filled in your tax form correctly and accurately, you just sign and date it, and send it back to the Inland Revenue in the envelope provided.

> **Tip**
> It is very important to keep a copy of your tax form. If there are any queries on it from your tax inspector it saves you having to hunt out – or work out – all the figures again.

Tax-free income

Some income is not taxable. And it includes the following:

- interest and proceeds from SAYE savings schemes, National Savings certificates, yearly plan, premium bonds
- first £70 of interest from the National Savings ordinary account
- proceeds from most regular premium life insurance policies that you have held for ten years or three quarters of their life (which ever is less)
- most grants and scholarships for education
- prizes from lotteries and winnings from betting
- small gifts from employers
- tax-free pensions and the £10 Christmas bonus
- tax-free social security benefits, including child benefit
- strike and unemployment pay from a trade union
- up to £30,000 in total of payments on leaving a job
- interest from delayed settlement of damage for personal injury
- interest on tax rebate.

Paying less tax – legally

The black economy is booming. The Inland Revenue reckon they lose £8 billion a year in uncollected tax from people who deal in cash and just don't tell the tax man that they made any money. These people are breaking the law.

Choose tax avoidance rather than tax evasion. It is just good common sense to pay as much as you have to, but no more, to the Inland Revenue.

WHERE TO GET HELP

If you are not sure about filling in your tax form, or are unclear on a specific point, the easiest and cheapest place to go for help is the Inland Revenue itself.

Your first call should be to your tax inspector, who should be able to get your file out and sort out the query, but don't expect him to be over-helpful.

Your tax inspector may be located in a different part of the country from you – which makes phoning expensive and a personal call impossible.

If that is the case, you can ask for your files to be sent to your nearest PAYE inquiry office, and a tax inspector there will meet you and attempt to sort out the problem.

If you have a complaint, or you and your tax inspector do not see eye to eye on some allowances or outgoings, there is an appeals procedure you can go through.

Once you get a notice of assessment you have thirty days in which to appeal.

Write to the tax inspector telling him you are going to appeal – and why – and also show him which figures you disagree with. Fill in a tax postponement application at the same time, and you won't have to pay the tax while you argue over the assessment. If you and the inspector fail to agree, your proposal will then go to the Appeal Commissioners, who will listen to both sides and decide the issue.

If your tax affairs are complicated you can use an accountant or financial adviser to deal with your tax form. Don't choose one at random. Ask around among your friends and colleagues to see who they use or try to find one who is used to dealing with people in the same line of business as you. That way they will know all the allowances you can claim.

But remember an accountant can be expensive. An accountant's charges will be £45 to £150 an hour so the more complicated your affairs are, the more you will have to pay.

However, you will often find that a good accountant saves you more than he or she charges.

Remember the Taxpayer's Charter. Along with many other government offices and companies, the Inland Revenue is giving its customers (remember when we all used to be called taxpayers) some rights. A copy of the charter comes with the introductory notes to your tax return. It is a good idea to read it so that you know what to expect from the Inland Revenue.

> **Tip**
> Your tax return has to be back with the Inland Revenue by 31 October. Don't leave it to the last minute because this is not a flexible date. If you don't get the return in by then – for whatever reason – you could be fined or overtaxed.

What if you are under- or overtaxed?

If you have filled in your form correctly and the Inland Revenue makes a mistake and charges you too little tax you could end up with a large and unwanted tax bill years later.

Don't panic. If the mistake is theirs and it isn't spotted within one tax year after the one in which the mistake was made, you may not have to pay the whole bill. It is called Official Error when it is their fault.

Providing you have been completely honest with the tax man and that your tax affairs are up-to-date and you don't owe the Inland Revenue any money then you will be let off completely if your income in the year of the assessment is less than £12,000.

Gross income between:
£12,001 and £14,500 you pay 25 per cent of what you owe
£14,501 and £18,500 you pay 50 per cent of what you owe
£18,501 and £22,000 you pay 75 per cent of what you owe
£22,001 and £32,000 you pay 90 per cent of what you owe
Over £32,000 you pay the lot.

If you are over sixty-five, receive a state pension or a widow's pension, these limits are all £3,300 higher, so you won't have anything to pay at all if your income is less than £15,300.

If you have overpaid more than £25 of tax, no matter whose fault it was, the Inland Revenue will pay interest on the money from one year after the tax year in which the mistake was made.

If you owe the Inland Revenue money and haven't paid up, you could be charged interest and penalties.

And if you don't tell your tax man about any new source of income that you have, regardless of whether or not you are sent a tax return, you could be penalized. The Inland Revenue is entitled to charge up to 100 per cent of the money owed as a fine, and ask for the tax, plus interest, as well.

8 Tax and Your Job

Checking your payslip · How a break in your work record affects your tax · Working abroad · Tax on expenses and perks

See also
- *Chapter 7, Understanding Income Tax*

Most people in this country work for someone else and that makes them employees. Whether you work for a huge multinational or are the only worker in a very small company, your tax treatment should be exactly the same. The tax rules for employees are quite clearly defined – you pay tax on your earnings and it is only in the area of expenses and fringe benefits that there is any leeway.

Employees are taxed under what's known as Schedule E. That means income tax is taken from your money under the PAYE system (Pay As You Earn) before you get it.

The nice thing about PAYE is that you really don't have to do anything about it. You fill in your tax form (if you get one), check your tax code when it is sent to you, and your employer and the tax man do the rest. For further details on filling in your tax form and understanding your tax code, read Chapter 7, Understanding Income Tax.

PAYE works like this. Your tax inspector tells your employer in January, February or March how much you are entitled to in allowances for the following tax year. Your employer then uses a set of tax tables to work out how much tax you are due to pay on the rest of your salary, deducts the money every month and sends it to the Inland Revenue.

You get your money net, and don't have to worry about any further tax bill on that cash.

At the end of every tax year your employer will give you a form called a P60. Don't lose it. The P60 tells you how much you have earned in that tax year, and how much tax and National Insurance you have paid on the money. You will need the form when you are filling in your tax return.

<table>
<tr><th>Employee's Name</th><th>Reference Number</th><th>Company Name</th><th>Date</th><th colspan="2">Period Number</th><th>Tax Code</th></tr>
<tr><td>Alison Mitchell</td><td>12345</td><td>Big Company plc</td><td>31-05-92</td><td colspan="2">02</td><td>290L</td></tr>
<tr><th>Basic Pay</th><th>Overtime</th><th>Adj</th><th>Tax</th><th>NI Contr</th><th>Other</th><th>Taxable Pay</th></tr>
<tr><td>£1,385.85</td><td>£324.70</td><td>£8.70</td><td>£352.00</td><td>£108.65</td><td>£0.00</td><td>£1,719.25</td></tr>
<tr><th rowspan="2">Pension</th><th colspan="3">Totals to Date this Year</th><th rowspan="2" colspan="2">Company Loan</th><th rowspan="2">Net Pay</th></tr>
<tr><th>Tax</th><th>NI</th><th>Pension</th></tr>
<tr><td>£69.06</td><td>£749.25</td><td>£219.80</td><td>£142.47</td><td colspan="2">£0.00</td><td>£1,189.54</td></tr>
</table>

Sample payslip

> **Tip**
> If you have to pay tax on your investment income, try not to let your tax man collect the money by changing your tax code. You can put off the pay-day by asking to be billed separately for the tax.

CHECKING YOUR PAYSLIP

The nice thing about being an employee is that you get paid regularly. Whether it is weekly or monthly, you can budget with ease because you know that money will be coming.

And there is nothing to beat that marvellous feeling of opening your pay-packet and taking out the cash, cheque or slip of paper that tells you how much has been paid into your account.

But don't discard the payslip. It explains the gap between what you earned and what you actually get. It may look like gobbledegook, but it's not. It is in code, quite a simple code actually, so if you want to check that you are being paid the correct money, become a code breaker.

All payslips contain the same basic information – though they are not all laid out in the same way as the one in the example above. The top line indicates who the payslip is for, and what period it covers. So it starts with the employee's name, a company reference number, the name of the company and the date – all very simple.

Then comes the period number. That is the tax week of the year. Week 1 is the first week of the tax year, which starts on 6 April, so the week of 31 May will be tax week 09. If an employee is paid monthly the system works in the same way, but the numbering represents months instead of weeks. So May is Period 02.

Then you'll find your tax code – that should tally with the tax code you agreed with the Inland Revenue before the start of the current tax year. For more detail on understanding your tax code see p. 74.

The next line details the work you have done in the period covered. First there is your basic pay – that doesn't include any holiday pay, overtime or bonus. They will be detailed separately.

There is also a box for any adjustments that have to be made. This would include any statutory sick pay, or payments from the company's own sick pay scheme or perhaps maternity pay if you are pregnant.

Unfortunately you can't stop at the end of this line! There are deductions which have to be taken off as well. You'll have National Insurance contributions taken off automatically if you earn over £54 a week and are under the state retirement age, and contributions to the company pension scheme, if you make any. There will also be a box for any other payments you have to pay to your firm, such as repaying a company loan or an advance on your salary.

The deductions are subtracted from all the additions and the resulting figure is your gross pay.

The tax will be taken off. The resulting figure is your net pay – what you actually get for the week or month.

The other boxes on your payslip will show running totals for the year – for National Insurance contributions, your gross pay, and your tax. If you don't understand any of the adjustments that have been made, ring your accounts department. Otherwise you won't know if you've been paid too much – or too little.

Tip

In May of each year there is an automatic adjustment to your pay to reflect the increases in personal allowances that were announced in the Budget. Your pay goes up in June.

INS AND OUTS OF WORKING

Very few people leave school, go to a job and work there continually until they retire. There are times when you will have a break, either to move to another job, have a baby, take a year or two off or have enforced leisure through unemployment or redundancy.

At every junction, it is important to know what your tax position is.

Starting work

When you start your first job after you leave school or college you will have to pay tax on anything you earn over £66 a week, or £287 a month, unless you are a married man getting the full married couple's tax-free allowance, in which case you can earn £100 a week or £430 a month tax-free.

To work out that figure, divide the personal or married couple's tax-free allowance by twelve if you are paid monthly, by fifty-two if you are paid weekly.

Single person's allowance of £3,445 divided by 12 = £287 a month
= £66 a week

These figures relate to the tax year 1992–3.

When your earnings breach this barrier you will start to pay tax at 20 per cent. Form P46 must be filled in by your employer. That will put you on to the PAYE system and give you a single person's basic code of 344L. That assumes you are entitled to the personal tax-free allowance and gives you £287 tax-free pay each month. If you have no other allowances, anything you earn over £287 a month is taxable.

If you start to pay tax midway through the tax year, say in the fifth month, your employer will use his tax tables to work out how much tax you are due to pay. He'll take the £287 of monthly allowances that you are entitled to, and multiply it by the five months that you haven't used the allowance. You will then get a lump sum of tax-free income against which you can offset your monthly salary.

You may not be asked to join the company pension scheme on day one of your new job – but if you are, think before you join.

Company pension schemes are generally very good news because the employer makes a contribution to the fund. And the more that is paid in, the more you will get when you retire.

However, nowadays very few people stick with the same company for the whole of their working life, so you may prefer to wait a few years before joining. And at the beginning of your working life you tend not to be paid very much, and there are probably plenty of other things to do with your cash. You might want to buy a house, go on nice holidays, save a bit for a rainy day or just have a year-long spending spree.

If you are thinking of joining the company scheme or starting up a personal pension plan, read Chapter 27, Pensions.

Once you start to pay tax you should fill in Form P15, which your employer will give you. This is a coding claim and it will make sure that you are being correctly tax coded. If the new tax code gives you higher tax-free allowances, you'll get a rebate on the tax you've paid.

Changing jobs

When you leave your job make sure your old employer gives you a P45. That form shows what your notice of coding is, how much you have earned in this tax year and how much tax you've paid. Give it to your new employer and he'll continue the system. That way you don't have to go on to an emergency coding where you may be paying too much tax. Although you would eventually get it all back, it is frustrating in the meantime to be paying so much.

Going back to work

Going back to work after a break of a few years – either through staying at home to look after young children, or because you have been unemployed – can be very different to starting work for the first time.

You will no doubt be earning more than the tax-free minimum of £66 a week or £287 a month. This time your employer doesn't give you the benefit of the lump-sum tax-free allowance. He has to start taxing you straight away.

You'll get the emergency code of 344L so anything you earn in a month above £287 will be taxed. Once you've filled in your P15 and have been given a tax code then any tax you have overpaid will be refunded.

Tip
Fill in your P15 coding claim and return it to your employer's PAYE office as soon as possible if you think you are being overtaxed. Otherwise you could be on an emergency code, and lining the coffers of the Revenue, however temporarily, for months.

Broken work record

Sickness

If you are off work because you are ill it will affect your tax position. By how much depends on who pays you when you are ill. Money from Statutory Sick Pay and/or money from your employer's own sick pay scheme is taxable.

Invalidity benefit, which takes over from statutory sick pay if you are still sick after six months, is tax-free.

If you have taken out a permanent health insurance policy to cover all or part of your earnings should you fall ill, you'll find that the money you get from it is not taxable until you have been receiving it for a whole tax year – that means you could have been getting it for up to twenty-three months. Thereafter it counts as investment income.

Pregnancy

State maternity benefits are tax-free. Statutory Maternity Pay, which is paid by your employer, is taxable. If it is paid while you are still working, it will be taxed under the PAYE system. If you've stopped working, tax will be deducted. In that case, you may be due a rebate because no tax-free allowances have been taken into account. Check with the Inland Revenue.

Redundancy

If you are made redundant some of the money you get will be tax-free, some of it taxable.

Anything that is part of your normal pay cheque, such as wages, pay in lieu of holiday, pay for working your notice and commission, will go through the PAYE system and be taxed accordingly.

Any additional redundancy payments or pay in lieu of notice (providing it is not a condition of service) are tax-free up to £30,000.

If your redundancy payment or golden handshake is over £30,000, the excess will be taxed at your top rate of tax.

Short-time working or laid off

If you are working less than you previously did because your employer has put you on short time or laid you off temporarily, then you will probably be due a tax rebate. It will be done automatically for you by your employer and refunded to you on your next pay-day.

Strike

If you stop work because you are on strike, you will probably be due a tax rebate when you return to work.

WORKING ABROAD

If you are sent overseas to work by your company you will not have to pay UK tax providing you are abroad for 365 days. That year does not have to coincide with the tax year or a calendar year.

The Inland Revenue does not have a heart of pure stone. You are allowed to come back to the UK for short periods during that 365-day period providing the total number of days spent in the UK is less than 91.

The length of these periods is very clearly defined:

- they have to be less than sixty-two days each
- they have to be short in relation to the time you have spent overseas.

The tax man uses the following formula: he assumes that you have gone abroad, come home for a visit and gone abroad again. He adds together:

A – the number of days you were abroad
B – the number of days you came home for
C – the number of days you were abroad again.

If B is less than one sixth of the total A + B + C, and less than sixty-two days, you will be OK. The 365-day period won't be seen as having been broken. If you come back to the UK for a second visit, the whole of the A + B + C period becomes A for the next sum.

To avoid paying UK income tax on the money you earn while you are abroad, it is absolutely crucial that you do your sums before planning your visits.

Example 1

Jim is posted to Saudi Arabia, and is entitled to 70 days leave.
He goes off to work for 80 days (A)
He comes home to visit for 10 days (B)

He goes back to Saudi
After 150 days (C) he wants to come home again
His first A + B + C sum shows that the 10-day visit was less than one sixth of the 240 day total

So his 80 + 10 + 150 all count towards his 365-day continuous period
On his second visit home he stays for 60 days
Then goes back for a further 100 days
His second A + B + C sum shows that 60 is again less than a sixth of 240 + 60 + 100
Altogether he has:

- not broken the 62-day rule
- not broken the one-sixth rule
- over 365 days of continuous out-of-the-country working

So he pays no UK tax on the money he earned.

Example 2

His friend Bill is not so smart. He too is entitled to 70 days leave.
He works for 80 days
He comes home for 50 days
Then he goes back to Saudi
After 130 days he wants to come home again
But Bill breaks the one-sixth rule because
80 + 50 + 130 = 260
260 divided by 6 is 43, which is seven less than the 50 days he came home for
So he starts his 365-day period on the day he returned to Saudi after his first trip home

Bill works for 130 days
He comes home for 20 days
He goes back for 120 days
This time he doesn't break the one-sixth rule so his 130 + 20 + 120 (270 days) all count towards making up his tax-free period. Sadly he is posted home again before he can reach the 365 days of continuous overseas work
Bill pays UK tax on his earnings.

Qualifying days. A day only qualifies if you spend the end of it abroad.

So the day you travel to your posting is a qualifying day, the day of your return isn't.

National Insurance contributions

If your employer is based in the UK, it is up to him to get a no-tax coding from the Inland Revenue. You can pay Class 1 National Insurance contributions for up to fifty-two weeks (for full details see p. 73), then you should consider paying the voluntary Class 3 contributions if you want to maintain your contribution record and qualify for the full state pension on retirement. (See Chapter 27, Pensions.)

EXPENSES AND PERKS

There's often more to a pay packet than money. You may qualify for expenses and fringe benefits. Some are taxable, some are not. As a rule of thumb, the more you earn the more benefits you are likely to pay tax on.

Expenses are not the same thing as fringe benefits – though if you are on the fiddle you may think that they are.

Expenses. Money paid 'wholly, exclusively and necessarily in the performance of the duties of your employment'. Either you pay it and your employer reimburses you, or he pays the expenses for you.

Fringe benefits. Extras given to you by your employer. These extras will not be cash payments, but they are usually part of your salary package designed to encourage employees to join and stay with a firm – for example a company car or private medical insurance.

EXPENSES

For an expense to be tax deductible it has to be allowable. That means that your tax inspector has to agree that you, or your employer, spent money on something that was necessary to your job, and only to be used for your job. That is where the phrase 'wholly, exclusively and necessarily' comes in.

For example, you might think that buying the monthly trade magazine was vital for you to keep up-to-date with developments in your profession or trade. Therefore it should be tax deductible. Wrong. It might be necessary to you, but not to your job – no tax relief.

Some expenses have already been agreed with the Inland Revenue. If you belong to a trade where you need specialist clothing or tools, you will probably find that a fixed deduction has already been worked out. And that fixed amount can be claimed every year, regardless of whether or not you spend the money. If you spend more, keep your receipts and claim what you spent. But remember this is not money that the Inland Revenue sends you, it is a tax-free allowance that you claim.

If you are not sure whether or not you can claim, ask your union or professional association, or ring your local PAYE office.

If an expense is not allowable you will be taxed on it if it is paid by your employer – either separately or through the PAYE system.

If an expense is allowable you will get tax relief – but in most cases your employer will still have to tell the Inland Revenue about it. He will use Form P9D if you earn under £8,500 a year, and Form P11D if you are a director or earn over £8,500.

There is an exception to this rule. Some employers have agreed a dispensation with the Inland Revenue so that they don't have to give full details of expenses.

Allowable expenses

- fees and subscriptions to professional bodies
- cleaning, maintaining and replacing protective clothes and uniforms; note – no ordinary clothes are allowable even if you buy them for work and never wear them outside of work (TV presenters hate this rule)
- repairing, maintaining and replacing special tools and musical instruments
- interest on loans to buy necessary equipment
- running costs of your car if it is necessary for work; you can claim a proportion of the costs if you use it for personal use as well
- proportion of the running costs of your home if you work from there
- wife travelling with husband on business – only if she has a use! – for example in business entertaining of overseas clients; this would also apply if the husband was accompanying the wife
- hotel and meal bills if you are travelling in the course of your job

Expenses and allowable expenses are a grey area when it comes to tax. One tax inspector might allow something, another may not. So if you are not satisfied with the way you have been treated, do something about it. First write to your own tax inspector asking him to justify his tax treatment of you, and if that does not clarify things you should follow the Inland Revenue complaints procedure (see p. 81).

FRINGE BENEFITS

How you are taxed on fringe benefits depends entirely on how much you earn. Regardless of how you see yourself, you are 'higher paid' in the eyes of the Inland Revenue if you earn over £8,500 a year. And that £8,500 a year includes your salary, the value of any perks you get and any expenses that are reimbursed to you for which your employer has no Inland Revenue dispensation.

Perks and the under-£8,500-a-year employees

If you earn less than £8,500 a year you are treated quite leniently by the Inland Revenue.

The tax due on anything that is given to you as a fringe benefit is based on the second-hand value of the gift.

If, for example, you get a £100 dress, the second-hand value could be seen as being £30.

If something is lent to you, such as a company car, it is not a gift so no tax is payable.

If something has no second-hand value, such as membership of a medical health insurance scheme or free chiropody at work, then no tax is due on it.

Tip

If you earn under £8,500 a year be careful that a new perk doesn't take you over it. A job that pays £6,500 a year plus a company car could be worth a lot less to you if your employer also offers you free medical insurance. The cost to the employer of the company car may not take your total salary package over £8,500, but when you add in the cost of the medical insurance as well, it could. You will not only have to pay tax on the cost of the medical insurance, you will also have to pay tax on the company car. Better by far to turn down the medical insurance.

Perks and the over-£8,500-a-year employees

You have a tougher time. There are some perks which are tax-free – see below – but the rest will be taxable. And you will be taxed on the cost of the fringe benefit to your employer.

So if you get a £100 dress and the cost to your employer is £100 then you will pay tax on the £100 at your top rate.

None the less, that is cheaper than paying £100 yourself, even if you are a 40 per cent taxpayer.

Don't turn up your nose at fringe benefits. If you want the benefit, they are much better value than extra cash in your pay packet.

Tax-free benefits

Whatever you earn, if your employer provides any of the following you won't pay tax on it:

- pension contributions
- life and sick-pay insurance premiums
- work place nurseries
- fees and subscriptions to recognized professional bodies, because they are tax deductible
- canteen meals, providing there are facilities for all members of staff (though not necessarily in the same canteen!)
- 15p a day luncheon vouchers
- overalls, protective clothing or uniforms essential for work
- specific removal expenses if you have to relocate or are taking a new job in a different part of the country
- loans of money if they qualify for tax relief
- rent-free or low-rent accommodation if you have to be there to do your job
- long-service awards (not money) for twenty years or more with the same firm, providing the gift doesn't cost more than £20 a year and you don't get more than one every ten years
- wedding or retirement gifts
- employees' outings and Christmas parties – up to £50 a head
- discounts on company products, providing the employer is not out of pocket
- awards for money-saving staff suggestions – up to £5,000
- free car-parking at work.

All other fringe benefits are taxable. The most common ones are worth looking at more closely.

Taxable benefits

Company car

The most widely available and popular fringe benefit is the company car. Whether you are offered a Porsche or a Mini, you'll find there is a huge gap between the amount of tax you pay, and the actual cost of running a car.

According to the AA, it costs £6,500 a year to keep a family saloon on the road – you'd have to have a very special car indeed to be paying anything like that in tax. This is one of the top perks you can get.

How much tax you pay depends on:

- the value of your car
- the engine size
- whether it is more than four years old
- how many business miles you do.

The value of your car. The tax tables divide cars into three values:

- up to £19,250
- £19,251 to £29,000
- over £29,000.

Tip
If the added extras on your car, such as air conditioning or up-market sound system, take your car into a higher bracket, have them added later. It is the purchase price of the car that counts.

Engine size. Cars are subdivided by engine size up to £19,250 only. If your car is worth less than that then it will be put into one of the following categories:

- 1400 cc or less
- 1401cc – 2000 cc
- over 2000 cc.

Over four years old. Cars which are over four years old are taxed at a lower rate than their newer brothers.

Mileage. If you travel more than 18,000 miles a year on business in your company car, your tax bill will be halved.

Company car tax, 1992–3

(*a*) Cars under four years old

Original market value	*Engine size*	*High business mileage (18,000 miles or more)*	*Average business mileage (2,501 to 17,999 miles)*	*Low business mileage (2,500 miles or less)*
£	cc	£	£	£
	0–1,400	1,070	2,140	3,210
Up to 19,250	1,401–2,000	1,385	2,770	4,155
	2,001 +	2,220	4,440	6,660
19,251 to 29,000	all	2,875	5,750	8,625
Over 29,000	all	4,650	9,300	13,950

(*b*) Cars over four years old

Original market value	*Engine size*	*High business mileage (18,000 miles or more)*	*Average business mileage (2,501 to 17,999 miles)*	*Low business mileage (2,500 miles or less)*
£	cc	£	£	£
	0–1,400	730	1,460	2,190
Up to 19,250	1,401–2,000	940	1,880	2,820
	2,001 +	1,490	2,980	4,470
19,251 to 29,000	all	1,935	3,870	5,805
Over 29,000	all	3,085	6,170	9,255

Tip

If you are not going to be using your car for a month – for whatever reason – hand the car and the keys back to your employer. That way you will cut the tax bill on your company car by a twelfth.

Petrol. You will have to pay tax on the perk if you get any petrol at all from your employer for private use – and that includes driving to and from work. It doesn't matter what quantity you get – a gallon a year or full reimbursement, the tax is the same.

Fuel charges 1992–3

Engine size cc	*High business mileage (18,000 miles or more)* £	*Average or low business mileage (0–17,999 miles)* £
Petrol		
0–1,400	250	500
1,401–2,000	315	630
2,001 +	470	940
Diesel		
0–2,000	230	460
2,001 +	295	590

The tax you have to pay relates this time to the engine size, not the price of the car, though you pay less tax if your car runs on diesel.

> **Tip**
> If you are offered the perk of free petrol, make sure it is worth your while. Drivers not doing much private mileage on the company petrol could end up paying more money in tax than they are saving in free fuel.

If you earn less than £8,500 a year, the company car is tax-free. Petrol is tax-free if your employer foots the petrol bill directly. If you pay it and are reimbursed then you will be taxed on the perk at the same rate as your higher-paid colleagues.

Running costs. If your employer pays directly for the general running costs of your car, such as servicing it, insuring it and so on then your tax position is unaffected. If you pay for them, claim the costs on your tax return as allowable expenses. If your car is used both for work and pleasure, you can claim a proportion of your costs.

Mileage allowances should be entered on your tax return as an expense allowance.

Pool cars. If more than one employee uses the car and no one takes it home regularly then it isn't a fringe benefit and no tax is due. The Inland Revenue rules are very strict here.

Cheap rate loans

Subsidized mortgages and loans are a major perk for bank, building society and insurance company employees. How much tax you pay depends on how much you earn.

Employees earning less than £8,500 (including the value of the perk) pay no tax on this fringe benefit.

Employees earning more than £8,500 a year get the subsidized mortgage tax-free if it is for buying their main home and comes within the £30,000 band of tax relief everyone is entitled to.

Other loans, for such things as season tickets, are valued differently. You have to pay tax on the difference between the rate of interest you pay to your employer, and the Inland Revenue's official rate. If that difference is less than £200 no tax is charged. But if you breach the £200 barrier you pay tax on the whole amount. Do your sums carefully before accepting a loan.

Watch out

Some perks are in the grey area of 'can be, can't be' taxed.

Removal expenses are one example. If you have to move house to join a firm, or relocate to transfer within the company, 'reasonable' expenses paid by your employer will be allowed. The definition of what is reasonable may be negotiated with your tax man and should include legal and estate agent's fees, the cost of the move, refitting carpets, even the cost of kitting out your children in new school uniforms. But I emphasize – negotiate with your tax man. You probably won't get everything. And keep all bills as evidence that you have spent the money. There is nothing that weakens a claimant's case so much as not having receipts for money they claim to have paid out.

Services such as a staff hairdresser or chiropodist are a tax-free perk when provided by your company. But if you spend your own money on outside professionals and are reimbursed by your company, then that money is taxable.

Medical insurance premiums for BUPA, PPP and so on are taxed at what they cost your employer. But there may still be a saving since your employer will benefit from a group discount.

Paying up

Your employer has to notify the Inland Revenue every year of the level of fringe benefits that staff receive. He will do this on a Form P11D. Try and get a copy of the information he has sent in on your perks, so that you can check that the figures are correct and that they tally with the figures you are filling in on your tax return.

If you have regular fringe benefits such as an over-£30,000 subsidized mortgage and a company car then you may have to pay the tax bill through having your tax code changed. That way the PAYE system will cope with collecting the extra tax.

Otherwise you will be billed separately by the Inland Revenue.

9 Tax and the Self-employed

Do you need an accountant? · Allowable business expenses · Your tax year · VAT

See also

- *Chapter 8, Tax and Your Job*

If you don't have a regular employer you may well be self-employed. And whether you are a workaholic or just do a little job on the side for a bit of spare cash you will have to come clean to the Inland Revenue.

Declare what you earn in a tax year, claim all the allowances and expenses that you are entitled to, and pay income tax on the rest.

If you decide to go freelance or turn yourself into a limited company, then the ball game changes. Your affairs do get much more complicated and at that stage, if you still haven't got an accountant, it is time to get one.

He or she is a professional. Looking after the accounts and financial affairs of a company like yours is his job, so he will expect to be properly paid for his services.

However, he should save you money by claiming every tax allowance that you are entitled to and making sure that your books are all in order. And, of course, his fee is tax deductible.

What is self-employed?

Self-employed people work for themselves. They don't have a regular employer, and don't pay income tax under the PAYE system. They are taxed under what's known as Schedule D. That means all of their income is paid to them gross – without tax having been deducted. That income is declared on an annual tax return, and the Inland Revenue sends them a bill for the tax due.

However, there are exceptions to this rule. You might think you are a self-employed nanny because you work for two or three mums a week, or a self-employed television PA because you are freelance. The Inland Revenue won't see you that way.

If you work for someone who tells you what to do, and supplies you with the tools of the trade, for example the baby's pram or the television studio, then the tax man considers you to be at the beck and call of the employer or employers and will insist that you pay PAYE.

You will not be classed as self-employed either, if you only work for one employer in a tax year. You might think you are freelance with a good year-long contract, but the Inland Revenue will see you as having a contract of employment and, thus, within the realm of PAYE.

You will only be self-employed if *you* dictate the terms on which you work. If there is any doubt, the Inland Revenue may try to prevent you being regarded as self-employed.

Do you need an accountant?

Whether or not you need a qualified accountant depends on two things:

- how complicated your business affairs are
- how financially competent you are.

Within bounds, most people could manage to organize their own financial affairs and those of their small business. But it takes time. Account books have to be kept up to date, staff paid, NI deducted and sent in, Inland Revenue tax forms filled up and, if your turnover is high enough, VAT to be collected and sent on.

So there comes a stage when you will need an accountant to help you. When that time comes depends on how financially competent you are. If you find all that sort of thing relatively easy – if time-consuming – you will soldier on alone for longer than someone who doesn't want to come to terms with it at all.

I believe you are better to get on with running the business you make money at, and leave accountancy to the professional accountant.

Switch from doing your own finances to using an accountant when you can more profitably use the time spent doing the books on running your own business.

Costs

An accountant costs money – and it is impossible to say how much. He, or she, will charge between £45 and £150 an hour, so if your accountant spends a long time on the financial affairs of your business, it will be expensive.

Most accountancy firms will give you an initial interview free of charge. During that time you will outline what you want done and you should, in return, be given some idea of the annual cost.

Many of the large firms of accountants have small provincial offices dotted around the country designed specifically to help small businesses, so don't assume that a large firm of accountants will always charge more than a small firm – though they may well do. Check before you appoint anyone.

What does influence the costs is the seniority of the accountant who is doing your books. If he is the senior partner, he will charge a lot more than if he is a newly qualified junior. If you only need a newly qualified junior, don't pay for the senior partner.

WORKING FOR YOURSELF

If you are self-employed, you won't send in a blue P1 tax return. You will be sent the more complicated brown P11. You will also have to send your business accounts to the Inland Revenue.

They can be as simple as a couple of pages. The first will detail what you earned in the tax year. The second will detail the expenses you incurred for your business which you think are tax deductible.

Without an accountant you will have to fight your own battle with the Inland Revenue on what expenses are allowable. Don't fall at the first hurdle. Fight hard for everything that you think should be allowed, and your tax man doesn't. There's a chance that he is wrong. If he goes ahead and assesses you, and you still feel that you're being hard done by, appeal (see p. 81).

Very small businesses

If your business is very small you only need to send a simple, three-line, summary of your tax affairs to the Inland Revenue. By 'very small', the Inland Revenue means you have a turnover of less than £10,000 a year. Businesses falling into that category need only declare the turnover, business expenses and net profits and that should satisfy the tax office. It should look a bit like this:

Summary of profits for year to 31 March 1992

turnover	£9,672
less business expenses	£3,068
net profit	£6,604

Expenses of your business

You should be allowed to claim:

- the basic running expenses of the business – that would include rent, heating, lighting, cleaning, phone bills, advertising, postage and stationery, and any newspapers, or specialist books or magazines you have to buy
- your accountant's fees and any bank charges on your business account
- Christmas party for the staff – up to £50 a head
- travel and accommodation expenses for business trips
- motor expenses
- interest on business loans
- money spent on hiring or leasing a car or other piece of business equipment
- insurance for the business
- some legal costs connected with the business.

If you work from home, you can claim a proportion of the household expenses. That would include the rent and poll tax/council tax, the phone and fax bill, heating, lighting and cleaning. To work out what proportion you claim, count the major rooms in your house. If there are eight rooms, and you work in one of them you could claim an eighth of all the bills. However you are well advised to claim only a ninth.

If the room you work in is solely used for your business, you could end up having to pay capital gains tax on a proportion of the profit you make when you sell the house. However, if the room you work in also doubles as the spare bedroom, or the dining-room, then you won't have any CGT liability. You will reinforce your argument that the room doubles as something else, however occasionally, by reducing the proportion of the household bills you claim.

> **Tip**
> Keep low the proportion you charge for rent, poll tax/council tax and water rates but you can justifiably increase the percentage for heat, light, power and particularly the telephone.

PAYING YOUR TAX

The nice thing about being self-employed is that you can delay paying your tax.

To delay as long as possible:

- choose your financial year end carefully
- check the rules on 'which profits are taxed when' carefully.

Financial year end

When you first become self-employed your initial reaction is to run your financial year to coincide with either the tax year (6 April to 5 April) or the calendar year (1 January to 31 December). That is not a good idea. Choose instead to end your financial year sometime soon after the end of the tax year – perhaps choosing 1 May to 30 April. That way you delay paying your tax.

Your income tax is due on 1 January and 1 July in the tax year following the one in which your financial year ends. So if your financial year ends on:

31 December 1991	tax is due on 1 January 1993 and 1 July 1993
30 April 1992	tax is due on 1 January 1994 and 1 July 1994

Which profits are taxed when?

To get to the accounting year dates that you want, your first year can either be shorter or longer than twelve months.

In that year you will be taxed on actual profits. That means you will be taxed on the profit you actually made from the day you started your business to the following 5 April (the start of the next tax year).

In practice the tax you pay will be based on a proportion of the profit shown in your first year's accounts (unless your accounting year coincides with the tax year).

In the third year you have a choice. You can opt to be taxed on actual profits. If you don't opt, you will automatically be taxed on the basis of the previous year's profits.

Actual-profits assessment means you pay tax in this tax year on the profit you actually made. Previous-year assessment means that you will pay tax in this tax year on the level of profits you made in the previous year.

Don't worry if you don't understand that concept.

Choose actual-profits assessment if you will make a lower profit in years two and three than you did in year one.

Allow previous-year assessment if you will make a higher profit in years two and three than you did in year one. This is generally the one most small businesses choose.

In subsequent years, the profits will be charged to tax on the basis of those made in the previous year. For example: profits for the year ending 30 April 1992 will be taxed in 1993–4.

NATIONAL INSURANCE

If you are self-employed you pay two types of National Insurance contributions. You pay:

Class 2 contributions regularly, by standing order usually, every month. The current 1992–3 level is £5.35 a week.

Class 4 contributions. They are based on your annual profits and are 6.3 per cent of any profit you make between £6,120 and £21,060. It is assessed and paid when you pay your income tax on your profits. You now receive tax relief on 50 per cent of these contributions.

VAT

Whether you are a self-employed single person or you run a small company, you will have to register for VAT as soon as your turnover reaches a certain level. In the 1992–3 tax year the registration threshold is reached when your turnover reaches, or is expected to reach, £36,600 a year. If it falls below £35,100 a year, you should de-register.

VAT is an indirect tax on consumers levied on nearly all goods and services. In most cases the rate of VAT is 17.5 per cent but on some goods, such as food, newspapers and children's clothes and shoes, the rate is 0 per cent. A few services, such as insurance, money lending, and any bills from your doctor, dentist or optician are exempt.

If you are self-employed and VAT registered you will have to add VAT to all your invoices. Any VAT that you pay for goods and services is reclaimable. The real bore about VAT is the tedious, time-consuming business of filling in the VAT return every three months. It is up to you to work out what you owe Customs and Excise, who run the system, and to send them a cheque on time.

To register for VAT, ring your local Customs and Excise office. They will send you all the necessary forms and a batch of pamphlets outlining quite clearly what you pay VAT on and what you can claim back.

They will also give you a date for registration. All invoices you send out after that date must include VAT.

Help cash flow

A recent change in the VAT rules to help small businesses allows you to opt for cash accounting if the turnover of the business is under £300,000 a year. That means you only pay the VAT to the Customs and Excise once you have received it from your creditor. If you don't opt for this system you have to pay the VAT on all invoices sent out during the three-month accounting period whether or not the bill has been paid. So using the cash-accounting system will help your cash flow.

Keep your VAT accounts up-to-date and accurate. Save all receipts for VAT you are claiming back, and a copy of all invoices for VAT you are charging. The VAT office is very stringent about this and within two years of your registering they will send an official to check your books. Use an accountant to set up and advise you on a book-keeping system which you could probably then maintain yourself. That way you will have proper records for the VAT man and you'll be able to use it to prepare annual accounts for the tax man.

And a word of warning – don't try to fiddle your VAT. The VAT man has right of entry to your house or business at any time to see and check your records.

Registering for VAT can save you money

If your line of business is freelance journalism, for example, you won't lose any money by being registered because the person commissioning your work won't care whether or not you charge VAT because he can claim back any VAT you charge. When you buy equipment such as a word processor or office furniture for your business, you will be able to reclaim the VAT you have paid.

Registering for VAT can lose you money

If you are in a trade such as building, gardening or painting and decorating, the customer won't want to pay the extra 17.5 per cent tax on your bill and may choose a rival who is not registered. In that case you may be better off, if you are a borderline case, keeping your business slightly smaller and turning down work rather than breaching the VAT registration barrier.

The VAT registration level goes up each year, and the new threshold is announced in the Budget.

EMPLOYING STAFF

Turning from being a one-man band into an employer is a major watershed. You not only expand your business, you expand your paperwork.

As an employer you are responsible for

- deducting tax through the PAYE system for your employees
- deducting National Insurance contributions from your employees and adding your own share
- keeping within all the employment legislation.

It is a minefield of rules and regulations and it is worthwhile getting on top of it from the start. If you don't you could spend a lot of time trying to put right your early mistakes.

Get in touch with the small business section at your local authority. They will have a basic pack of general information designed to set you on the right lines. And if you have specific questions or worries they can point you in the direction of specialists who will help you. They often run courses to help you with setting up in business.

Ask your local Inland Revenue PAYE office and your local DSS office to send you their leaflets on employing staff:

Inland Revenue
IR 28, 'Starting in Business'
IR 56, 'Tax: Employed or Self-Employed'

Other helpful Inland Revenue leaflets include:
IR 57, 'Thinking of Working for Yourself'
CGT 11, 'CGT – The Small Businessman'
IR 104, 'Simple Tax Accounts'

DSS
NP 15, 'Employer Guide to NI Contributions'
NI 227, 'Employer Guide to Statutory Sick Pay'
NI 257, 'Employer Guide to Statutory Maternity Pay'

SPARE-TIME INCOME

If you have a hobby that's turning profitable or you work in a bar at night, or you do some spare-time work from home in addition to your normal day job, then it will alter your tax position.

Of course you have to tell the Inland Revenue about it. And to do that, fill in the details of your spare-time job on your tax form. If you don't get a tax form, write to your local tax office giving them the name and address of your employer (if you have one) and notifying them of your new sources of income. Keep a copy of your letter.

If you don't you could be fined when the Inland Revenue do find out about it (and they will) and interest will be payable on the tax due.

AVOID A TAX BLITZ

If the Inland Revenue have good reason to believe you are not paying as much as you should be – that is, you are evading tax – they will come down on you like a ton of bricks.

A tax blitz is the last thing any small business needs. Your accountant will have to supply the Inland Revenue with all sorts of documents and details and it is very time-consuming. Anything which takes up your accountant's time will cost you money. So be honest – in the long run it will probably save you money.

Finally, the last thing that the tax man, the VAT man or any accountant wants is to sort out a mess – so don't get into one.

10 Tax and Your Investments

Income tax and capital gains tax – what you pay it on and how to avoid it · Which type of investments suit which type of taxpayer

See also

• *Chapter 3, Building Your Savings*

Most people have some sort of investment, or unearned, income. You don't need to have a large portfolio of shares, or a stake in the gilt market to qualify. A single building society or bank account will do.

Any income that doesn't come from your job, your business or your pension is classed as investment income – and taxed as such.

There are two types of tax you can pay on this money:

- income tax
- capital gains tax (CGT).

You pay income tax if the money comes as income from your capital, that is interest on a savings account, or perhaps a dividend from a share or unit trust. And you pay the tax at your top rate. So if you are earning enough to pay tax at more than the basic rate on the top slice of your income, that is at 40 per cent, then that is the rate of tax you will pay on your investment income.

You pay CGT on gains you make with your capital. If you sell a share at more than you paid for it, or your valuable antique table has trebled in value, then the gain you make when you sell will be liable to CGT. However, you don't pay CGT on every penny of gain you make. The first tranche of gain that you make in any one tax year is tax-free. The level usually goes up annually – though in the 1988 Budget it was reduced to compensate for the income tax cuts – and at the moment it stands at £5,800. You can make a capital gains profit of £5,800 in the 1992–3 tax year before you start to pay tax. Again, you pay tax at your top rate.

INCOME TAX

Let's start with the basics.

Bank and building society accounts

The interest on these accounts is subject to income tax. Tax is regarded as having been deducted at the basic rate, but it can be claimed back. Non-taxpayers should fill in a form at their local bank or building society branch.

Higher-rate taxpayers should not have much money in one of these accounts because they will have to pay extra tax to the Inland Revenue, to make up the difference between 25 per cent and their actual top rate of tax, 40 per cent. That makes the interest rate – or return on their capital – very low indeed; they should be able to do better elsewhere. For high-rate taxpayers, bank and building society accounts should only be used for easy-access money.

National Savings accounts (see p. 14)

Few National Savings accounts have tax deducted from the interest at source. This interest is usually paid:

- tax free – in which case no tax is due on the interest no matter what your tax position is
- gross – the interest is subject to tax if you are a taxpayer. Non-taxpayers keep the interest in full. If you are a taxpayer you will have to declare the interest on your tax form and pay the tax, at your top rate.

National Savings tax-free accounts include:

- savings certificates
- yearly plan.

National Savings 'paid gross' accounts include:

- the investment account
- income bonds
- capital bonds
- children's bonus bonds
- guaranteed growth bonds.

The hybrid National Savings ordinary account pays a miserly rate of interest, but allows the first £70 of annual interest tax-free; thereafter interest is taxed at your top rate. This account should only be used by high-rate taxpayers.

Shares and unit trusts (see p. 32)

The income you get from shares and unit trusts is known as dividends. The money (which comes from a division of the company's profits) will be sent to you, usually twice a year, in the form of a cheque, or it can be paid directly into one of your bank or building society accounts. Dividends you receive will already have had tax deducted at the basic rate of 25 per cent. The statement attached to the cheque will detail how much tax has been paid and if you are a non-taxpayer or a 20 per cent taxpayer you can claim all or some of this money back. Send the voucher and a covering tax claim to your tax inspector – or your local tax office if you don't have an inspector – and he should send you a cheque. If you are a 40 per cent taxpayer, you will have to pay extra.

Gilts (see p. 27)

If you invest in gilt-edged securities, you will get interest sent to you twice a year. If you are a taxpayer you will have to pay tax on that interest. How you pay tax depends, in the first instance, on how you bought the gilts. If you bought them through the Post Office Register – directly from the National Savings office at Lytham St Annes – the interest will be paid gross, that is without tax having been deducted. So you will have to declare that interest on your tax form. If you bought through a stockbroker, then tax at 25 per cent will have been deducted from your interest before you get it. Non-taxpayers can claim this tax back, high-rate taxpayers will have to pay extra, 20 per cent taxpayers can reclaim some of it.

All income from unit trusts, shares and gilts should be entered on your tax form. If you have paid the tax you won't have to pay it again, but if you haven't or you pay tax at more than 25 per cent, there will be a charge and your tax inspector will either change your tax code the following year, or ask you for the money.

PEP (see p. 46)

You will not have to pay any income tax on the income from the money invested in a PEP. The tax that is automatically deducted from dividends will be reclaimed by your PEP fund manager and reinvested in the PEP. There is no capital gains tax to pay on the capital profits either.

CAPITAL GAINS TAX

Bank and building society and National Savings accounts only go up in value by adding interest. They don't show a capital gain, so no CGT will ever be payable on these accounts.

Shares and unit trusts

If you sell a share or unit trust for more than you paid for it you will be liable to CGT on the difference. The first £5,800 of gains for the year (and that figure usually rises annually) are CGT-free; thereafter you pay CGT. However, if you make any losses – that is you sell your shares for less than you paid for them – you can offset these losses against your profits. If you are really hopeless at investing and make a loss over the year, don't expect the Inland Revenue to pay you 25 per cent or even 40 per cent of that money. There is no such thing as reverse CGT. However, you can carry forward any losses to be offset against capital gains in future years.

It is worthwhile checking your portfolio towards the end of the tax year. If your shares are showing a total profit edging towards the CGT-free limit it might be worthwhile selling some of them to realize that profit. That way you may keep your following year's profit also within the CGT-free limit. After all you can always buy back the shares if you think they will go still higher.

As the example below shows, the savings can be quite substantial.

Example 1

Year 1
You buy 5,000 shares in Big Company plc for 150p each
In March the shares are worth 250p
Your potential profit is 5,000 x 100p (250p – 150p) = £5,000
Sell the shares and buy them back

Year 2
The shares rise to 350p
Your potential profit is another £5,000
You sell the shares and take your profit
Profit over two years £10,000
Tax – none
(There will be a dealing cost of around £300 for buying and selling the shares.)

Example 2

Year 1
You buy 5,000 shares in Big Company plc for 150p each
In March the shares are worth 250p
You do nothing
Year 2
The shares rise to 350p
You sell the shares and take your profit
Profit over two years is £10,000
CGT bill is £1,680 (£10,000 – £5,800 = £4,200; 40 per cent of £4,200 = £1,680)
(assuming you pay tax at 40 per cent on your top slice of earnings)

These examples do not include indexation.

Indexation

You are allowed another concession on your capital gain – and that is you can take inflation into account. Your asset, whether it is shares, gilts, antiques or whatever, is allowed to increase in value by the rate of inflation every year. The gain above that is then subject to CGT. These rules came into being in 1982. Anything you bought before that date has to be dealt with differently (see below).

If you are getting into the realms of applying the indexation allowance to your capital gains, your affairs are probably complicated enough to need the help of a qualified accountant or professional adviser.

But if you want to go it alone this is the indexation allowance formula. The allowance is equal to the cost of the item or the market value in the month that you bought it (including any expenses incurred in buying it) multiplied by (the retail price index in the month in which the asset was sold minus the RPI of the month in which the asset was bought divided by the RPI in the month in which the asset was bought or 31 March 1982)!

Example

If you are selling shares worth £10,000 and the RPI was 400 when you bought and 500 when you sold, the indexation will be:

$$£10{,}000 \times \frac{500 - 400}{400} = £2{,}500$$

This £2,500 is then added to the original cost of the item when calculating the capital gain (or loss) arising.

To find out what the RPI was in the month you bought the item, or shares, and in the month which you sold, ring your tax inspector and ask him.

Before 1982

If you are selling something that was bought before 1982 the Inland Revenue allows you to discount any profit you made in the years between the date you bought your item and 31 March 1982. This new rule, introduced in the 1988 Budget, will help considerably anyone who bought something before the inflationary years of the 1970s and is selling it now.

If the reason your second house went up in value three times was solely that inflation was running at over 20 per cent a year in the seventies, the Chancellor does not see that you should be penalized.

You will have to find, and agree with the Inland Revenue, the 31 March 1982 value of your item and index the profits from then until you are selling, and that should reduce your profit margin.

Valuable items

Buying paintings, antiques, expensive jewellery, gold coins or wine can be an enjoyable hobby. If it turns into a money-spinner, then the tax man will expect his share. You will have to pay CGT on any profit you make in the same way as if it was a share.

You deduct what you paid for your asset from what you sold it for, take off any expenses such as legal fees or commission, calculate the indexation allowance and declare the result on your tax form.

Your £5,800 allowance counts against this sort of gain in the same way as against a profit from shares or gilts, but remember you only have one allowance. If you use it up on your shares, you will have nothing left to set against your asset sales.

But there is another allowance here. If the item you sell makes a profit of less than £6,000, the Inland Revenue will ignore it. Don't think you can fool the tax man by selling half a dozen antique chairs separately to the same person so that the gain on each stays under £6,000. The Inland Revenue is not stupid and would disallow that sort of flagrant disregard for their rules.

Apart from the 'under £6,000' rule there are other categories of gains that are CGT-free. These include:

- the proceeds of a life insurance policy
- timber, providing you are not running a lumber business
- the proceeds from your pension
- the proceeds from selling a medal for valour – providing you didn't buy the VC or George Cross in the first place
- anything that is not expected to last more than fifty years such as a yacht or a racehorse
- gifts to charities.

Second homes

When you sell your home – that is your 'principal private residence' – you do not pay CGT on the profit that you make. However, if you have two homes, they are not both exempt. You have to pay CGT on any profit you make when you sell the one that is not your main home.

It is up to you to choose which of your properties is your main home. It doesn't have to be the one you live in most, nor the one where you get tax relief on the mortgage, but you would have to live in it part of the time. So if you are nominating one of your houses to be your main home, choose the one that is likely to show the larger capital gain.

You have to write to your tax man within two years of buying your second home telling him which home is your main one. Until the property is sold, you can change your mind at any time and backdate that decision by two years. A married couple will both have to sign the letter unless both homes are owned by one partner. If you don't tell the tax man, he will tell you which is your main home. That's the theory; it doesn't always work in practice.

If you live in rented accommodation that is tied to your job – perhaps you are a vicar, a hotel manager or a caretaker – and you buy a home which you intend to move into eventually, write and tell your tax man that the house you've bought is your main home. If you are reading this more than two years after buying the home and you haven't told the Inland Revenue, don't panic. Tell them now and ask for the nomination to be backdated to the date you bought your house. Your request may be accepted.

Business expansion scheme

The BES system was set up to encourage those with money to invest in those without, or needing more, but it is being phased out in December 1993. Small companies struggling to survive don't need the swingeing interest-rate payments that banks would ask on a loan; instead private investors can put up the money in return for a share of the business. The incentive is huge tax relief to the people who lend the money.

Until December 1993, you can invest between £500 and £40,000 in each tax year, and you can get tax relief on the money at your top rate of tax. A 40 per cent taxpayer wanting to invest £10,000 puts the £10,000 into the company and then claims £4,000 back from the Inland Revenue. So he only risks £6,000 but has a £10,000 stake in the company. In the days when the top tax rate was 60 per cent, a high-rate taxpayer could reclaim as much as £6,000 of his £10,000 investment, so by reducing the tax rate, the Chancellor also reduced the tax rebate!

The down side is the risk. These companies, because they are small and new, often go under, taking your money with them.

If you don't know any companies looking for money, or you want to spread your risk by investing in several companies, you can go through one of the BES funds. They work rather like unit trusts. They pool investors' money and spread it among several new companies which need investment cash to grow. But the risk can be high.

There are several rules you have to obey to qualify for the full tax relief:

- you can't own more than 30 per cent of one company
- you mustn't be a paid director or employee of the company or relation of a director
- you have to hold the shares for at least five years.

And the company has to fulfil certain criteria too:

- it has to be a UK trading company and UK resident
- it has to be unquoted
- it cannot raise more than £500,000 in any one tax year, unless its business is ship chartering or renting residential property to the private sector, when the limit is £5 million.

And to close some of the loopholes that live-wires in the City took advantage of, the companies cannot:

- hold commodities as investments, for example, wine or antiques
- hold more than half their assets in land or property
- be in the following businesses – banking, farming, insurance, leasing or hiring, share dealing, accountancy or legal services or property dealing.

But if you are out to make a genuine investment, and are not looking for a tax dodge, you should have no bother with these rules and regulations. If you put money into a BES scheme you will need a certificate from the company that shows that the Inland Revenue conditions are met – it should be Form BES 3 or Form BES 5. You send that, along with a covering letter, to your tax inspector. Don't delay. If you don't claim within two years of the end of the tax year in which you make the investment, you lose your right to tax relief.

BES schemes can be complicated, so get expert tax advice before parting with your money to make sure that you will be able to qualify for the tax relief.

No new money can be put into BES schemes after 31 December 1993.

SAVING TAX

If you get your savings into the right place you can save a lot of tax.

The non-taxpayer

Your money should be invested where interest is paid gross or where tax can be claimed back.

> **Tip**
> Don't choose an investment for capital gain if an investment for income will do. You are not exempt from CGT – though if you make enough to come into the CGT bracket it is likely that you will be paying income tax.

The basic-rate taxpayer

You can put your savings anywhere you like; keep your first tranche of money in a building society or bank account and open a TESSA.

If you are prepared to take a bit of a risk, make use of the fact that you are unlikely to pay capital gains tax. Everyone is allowed to make capital gains of up to £5,800 (in the 1992–3 tax year) CGT-free. So, instead of keeping all your money on deposit, try going for unit trusts or shares that might go up in price during the year (though remember that they might not). That way you can make a capital gain which is tax-free. But remember shares and unit trusts go down as well, so don't gamble what you can't afford.

If you don't mind locking your money up for a year or two try buying a PEP or some gilts. Choose a gilt that is standing at less than £100 for £100 of stock, but is due to be redeemed in the next few years. You can hold it to redemption and the gain you will make is tax-free. If you don't need the income, choose a low coupon gilt, that is a gilt that offers very little interest. They are cheaper to buy but offer a capital gain (which is what you want) rather than six-monthly interest (which is what you don't want).

High-rate taxpayers

Avoid putting too much into bank and building society accounts. The interest, after top rates of tax, is so low for you that it is not worth bothering about. But remember, of course, to open a TESSA.

You want accounts that are tax-free such as National Savings certificates; put a lump sum into the National Savings ordinary account where the first £70 of interest is tax-free.

Put the maximum of £6,000 into a PEP every year (and remember husbands and wives can put £6,000 in each).

Switch some of your income into capital gains and take up to £5,800 tax-free every year. Buy gilts, unit trusts and shares and consider investing in a business expansion scheme. But remember, you put your money at risk by switching from boring savings accounts into investments linked to the stock market and BES schemes. A tax-free gain is not guaranteed.

Helpful IR leaflets on CGT
CGT 4, 'CGT – Owner-occupied Houses'
CGT 11, 'CGT – The Small Businessman'
CGT 13, 'CGT – The Indexation Allowance'

11 Inheritance Tax

Who pays inheritance tax · How to avoid it

See also
- *Chapter 6, Life Insurance*

No matter how much or how little you have – you can't take it with you. When you go, and we all do, you have to leave your worldly wealth behind. Whether you leave it to your nearest and dearest, Millwall Football Club or the tax man is up to you.

But if you want to keep to a bare minimum the amount the Inland Revenue get their sticky fingers on, do some tax planning now.

What you are trying to avoid is inheritance tax. That's the tax that has to be paid if you leave more than £150,000 to anyone other than your spouse or a charity. Any taxable gifts made in the previous seven years will also be included. And with the surge in house prices over the past twenty years in most parts of the country, many people who would never have dreamt they would be so wealthy are now in the inheritance tax bracket.

WHAT IS INHERITANCE TAX?

When you die all your assets and any gifts made over the past seven years will be added up and if they come to more than £150,000 the excess will be taxed at 40 per cent.

Under £150,000	no tax
Over £150,000	40 per cent tax

To avoid paying tax the easiest thing to do is to give your money away while you are alive. To stop you handing over the estate on your death-bed, the tax rules insist that you live seven years after you make any gift, with some exceptions. You must also give away any property, cash or investments absolutely. You can't, for example, continue to live in a house that you have 'given' to your daughter. The seven years will not start to run until you have moved out. Nor could you 'give' your ICI shares to a grandchild, but keep receiving the dividend yourself.

The seven-year rule on gifts

If you die within seven years of making the gift, and you are leaving over £150,000 (including taxable gifts given in the last seven years), your estate will be liable for some tax on the money.

The Inland Revenue uses the following scale:

Years between gift and death	*Percentage of tax payable*
Up to 3 years	100
3–4 years	80
4–5 years	60
5–6 years	40
6–7 years	20
More than 7 years	0

If you live for seven years after making the gift, it will be free of all inheritance tax provided the money went to:

- an individual
- an accumulation and maintenance trust
- a trust for a disabled person
- an interest in possession trust.

If the gifts total more than £150,000 and do not come within the above four categories then inheritance tax has to be paid when the gift is made. But only at half rate – that is 20 per cent – and subject again to the seven-year rule.

Inheritance-tax-free gifts

There are some exemptions and reliefs. No inheritance tax need be paid if the gift is:

- between a husband and wife
- to a UK charity
- to a UK political party (which must have two MPs)
- totals no more than £3,000 a year; there is a one-year carry forward on this one, so if you don't give away £3,000 one year you can give

away up to £6,000 the next, but you can't carry the unused £3,000 forward further than that

- not more than £250 a year to as many people as you like provided it is not the same person or people as the ones who are getting the £3,000; these small gifts don't count in running up the £3,000 total
- normal expenditure out of income; these have to be regular gifts, though not necessarily to the same person
- a wedding present: if you are the parent of the bride or groom you can give up to £5,000; if you are the grandparent or great-grandparent you can give £2,500, otherwise you can only give £1,000
- part of a divorce settlement
- to support your mother or mother-in-law if she is widowed, separated or divorced or any other old or infirm relative
- certain works of art and woodlands
- 100 per cent / 50 per cent relief on agricultural land or business assets and unquoted shares; if you are a member of Lloyd's, deposits, reserves.

Inheritance tax planning

There are steps you can take now so that less inheritance tax is paid on your estate. But remember it is not you that pays the tax, it is the people you leave your money to. So don't inhibit your lifestyle just so that they will get more of your money. Anything you spend bears no inheritance tax!

- If you are rich enough, use up your exemptions every year – give £3,000 away and as many little gifts of under £250 as you like.
- Take out a life insurance policy in a trust for your children or grandchildren (for further details see Chapter 6, Life Insurance). Providing you can pay the premiums out of income – and don't have to dip into your capital to do it – this will come under the 'normal expenditure out of income' category. You have to 'assign' the policy to trustees so that it doesn't come into your estate when you die. If it did you would compound the inheritance tax problem because the estate and therefore the tax bill would be larger. It is a very easy process to 'assign' the policy – your professional adviser, solicitor or the life insurance company will help you with the details.
- If you can afford it, give it away during your life and live for seven years.
- Remember that your spouse has the same exemptions and reliefs,

including the £150,000 nil rate band and the £3,000 annual gift exemption.

- If you are past your first flush of youth, make sure your will is correctly drawn. Don't just leave everything to your surviving spouse even though no tax would be payable. They could leave a large inheritance tax bill behind when they die. Pass part of your estate – how much depends on how wealthy you are – directly to your children and if it is under £150,000 it will pass to them inheritance tax free.

Example 1

John Smith dies leaving £300,000 to his wife Ethel
No inheritance tax is paid
One month later Ethel dies
She leaves £300,000 to her son Jim
Inheritance tax bill is:

£150,000	no tax
£150,000	40 per cent tax
Money to Inland Revenue	£60,000

Example 2

John Smith dies leaving £300,000
He bequeaths £150,000 to his wife Ethel and £150,000 to his son Jim
No inheritance tax is paid
One month later Ethel dies
She leaves £150,000 to Jim
Inheritance tax bill is:

£150,000	no tax
Money to the Inland Revenue:	£0,000
So the total saving is	£60,000

Of course Ethel might have lived another twenty years so it is important not to leave too much to your children at the expense of your surviving spouse.

What you can't do is give something away, but retain an interest in it. For example you couldn't give your home to your children, but continue to live in it. If you want to give them your home to avoid it coming into your estate, you would have to pay your new landlords – your children

– a full market rent. Or you would have to move out and live elsewhere . . . for another seven years at least.

INSURING AGAINST AN INHERITANCE TAX BILL

If you don't want to give your money away while you're alive, or you worry about not living for the necessary seven years after you've done the decent thing – then you could use some insurance.

If you want to cover the potential inheritance tax that would have to be paid on a gift if you didn't live for seven years after giving it, buy decreasing term insurance (see Chapter 6).

If you want to cover the potential inheritance tax bill that will be due on your estate, you will need a whole life policy. If you are a couple, take out a joint-life, second-death policy. It will pay up on the death of the last survivor – as the insurance companies so quaintly term it – and it is cheaper than taking out two separate single life policies.

What tax-planning steps you take to minimize the amount your beneficiaries have to pay to the Inland Revenue depends on your age.

Over seventies

If you want to cover the liability to inheritance tax on your estate when you die, you'll have to go for a 'back to back'. It is a complicated concept but if you are over seventy and have an estate worth more than £250,000 it is worthwhile taking the time and trouble to understand it.

All you really need is a life insurance policy which will pay out a lump sum on your death equal to the projected inheritance tax liability on your estate. The problem is that, because of your age, that policy would be very expensive. The premium is paid monthly or annually so if you confound all the statistics and live until you are ninety-five – still paying the regular life insurance premiums – you will pay in much more than the expected pay out.

To avoid this you take out an annuity at the same time. An annuity is a life insurance policy in reverse. You pay in a lump sum, the insurance company pays out a monthly income to you.

Now the clever thing is that the annuity will pay the life insurance premiums and give you an additional after-tax income of around 5 per cent of the money. And both keep going until your death. Of course there's a catch. You have to put up the lump sum for the annuity. But that lump sum would probably have been left to your heirs when you died and would have been liable for inheritance tax. So it is not as big a sum for your heirs to lose as it might seem at first sight.

It is the only way to give and receive at the same time.

All of these policies are written in trust for your beneficiaries and are on the lives of both partners if the couple are married and pay up on the death of the survivor. Before being accepted by the insurance company as an agreed risk, you would have to pass a medical. You don't have to be fit enough to run a marathon, only well enough for someone of your age.

Under seventies

Because you are younger you may be able to afford life insurance on its own. What you will need is a high level of life cover quickly so you should go for a low-cost, or an appropriate unit-linked policy, rather than a full with-profits policy (see p. 56). You don't need the extra life cover. Low-cost and unit-linked policies offer that high level of cover more quickly but the down side to that is that they don't then go on to build up in value in the way that with-profits policies do.

This area is a complex one – not least because there could be a change in the inheritance tax rules. If your estate is so large that you are thinking of life policies to cover the inheritance tax liability, or want to set up trusts for your heirs, you would be well advised to get help from a financial adviser who is a specialist in inheritance tax planning.

12 Tax Checklist

Tips for avoiding tax · Tax mistakes · How to complain

For fuller information on all the points in the checklist, read the previous chapters in the section on tax (Chapters 7–11).

There really is only one golden rule when it comes to tax affairs – be honest with the Inland Revenue. Once you start dipping a toe into the black economy – however tentatively – you are storing up trouble for yourself. In the end the tax man will probably find out about the income you are keeping secret. And once that happens you are in for a rocky ride. If you are honest and declare everything up front, you will be dealt with fairly. There is a taxpayers' charter to guarantee your rights. By the time you've offset your costs against your profit you will more than likely find that your tax bill is not as harsh as you expected.

None the less there are plenty of honest ways of saving tax – and everyone should take advantage of them.

WAYS OF SAVING TAX

- Check that you are claiming all the allowances you are entitled to.
- Check that your tax code is correct.
- Make sure a wife claims the married couple's allowance if she is a taxpayer and her husband is not; or if she pays tax at a higher rate.
- Tell the Inland Revenue if you take out a personal pension. You will get tax relief on the contributions.
- Tell the Inland Revenue if you get married.
- If you invest in shares, try to keep your profits below £5,800 a year and you won't have to pay capital gains tax.
- If you find you have been overpaying tax, you can claim back the money you've overpaid during the previous six years.
- If you are a single parent, claim the additional personal allowance.

- If you work from home, remember to claim a percentage of household expenses.
- If you are buying a house for just over £30,000, divide the price into the cost of the house, and the cost of the fixtures and fittings. If that brings the house price below £30,000 you will save over £300 in stamp duty.
- If you earn under £8,500 a year, receiving fringe benefits may be better than a wage rise but watch that the benefit plus your salary doesn't breach £8,500.
- If you give money regularly to a charity, use a covenant or Gift Aid.
- If you are a non-taxpayer and have a bank or building society account, remember to sign the form exempting you from tax.
- If you are a high-rate taxpayer, go for capital growth rather than income and remember to take £5,800 of profits every year.
- If you qualify for the age allowance, watch where you keep your savings. Borderline cases can be hard hit if they choose a building society instead of the tax-free National Savings certificates.
- In the last ten to fifteen years of your working life, put as much of your spare cash as you can into your pension plan via AVCs (additional voluntary contributions).
- If you have more than one home, remember to tell the tax man which one you choose as your 'main home'.
- If your heirs will have a large inheritance tax bill when you die, consider giving away some of your money now.
- If your company has a share option scheme, join if you can. It is a no-lose situation.
- If you have a child going through college and you have set up a covenant to help with the parental contribution, don't cancel it. You won't be able to get tax relief on a new one. The same applies to a grandparent helping with school fees through a covenant.
- If you are posted abroad, be careful about how long you come home for. If you break the rules you could get an unexpected UK tax bill.
- For tax-free savings use a personal equity plan. But remember: there will be an element of risk.
- Remember that the tax man can be your friend – if you have a problem give him a ring or write to him and he may be able to sort it out for you. If you want to see your Inland Revenue inspector, and he does not work in your area, you can ask for your tax papers to be sent to your local PAYE office, and an inspector there will go over the details with you.

GIVING TO CHARITY

There are three tax-efficient ways of giving money to charity:

- by deed of covenant
- through payroll deduction
- by Gift Aid.

Deed of covenant

A covenant is a very old-fashioned term for a very up-to-date method of saving tax. Until April 1988 you could use tax-efficient covenants to help your child through college or perhaps help your god- or grandchildren with school fees. No more. Tax relief on new covenants is only given if the recipient is a registered charity.

A covenant is an agreement to make regular payments, for at least three years, to someone else – nowadays a charity.

The person paying the money pays net, and the charity can claim back the tax that has already been paid.

So if you donate £75 out of your net income, the charity can claim back the £25 that you have already paid in tax.

If you pay tax at 40 per cent, the charity can still only claim relief at 25 per cent but you can claim the additional relief from the tax man yourself. Non-taxpayers would have to pay the tax to the Inland Revenue.

How covenants work

To pay money through a covenant, you have to follow certain basic steps and fill in specific Inland Revenue forms. They are much simplified nowadays. If you are giving money to a large charity you will probably find that they have covenant forms they can send you.

In principle this is how a covenant works:

You agree to give the charity	£100 a year
You actually hand over	£75 a year (£100 less basic-rate tax at 25 per cent)
The charity accepts the	£75 from you
And claims from the Inland Revenue	£25
You pay	£75
The Inland Revenue pays	£25
The charity gets	£100

You can covenant any amount of money you like and, providing you have paid income tax on it, the charity can claim that tax back.

However, you don't all stand round in a circle with bundles of cash once a year. You have to fill in the appropriate Inland Revenue forms and it does take some time for the Inland Revenue to send its cheque. However, once the covenant has been accepted and the regular payments started it should all work smoothly and quickly.

> **Tip**
> Don't make the first payment until after the covenant has been signed and dated, or the first payment won't be allowed. Covenants cannot be backdated to cover money already paid.

Payroll deductions

Since the introduction of payroll deduction schemes in the 1986 Budget, over 2,000 company schemes have been set up.

An employee can now give up to £480 a year or £40 a month to the charity of his or her choice. The money is taken automatically from the employee's wages before tax is deducted and is given, through agencies, to the charity or charities chosen by the employee.

Gift Aid

Over £150 million has been given to charities this way by companies or individuals wanting to make one-off payments. By claiming back the tax the charities have increased this figure to £200 million.

A person or company can make a one-off payment of at least £400 (net of basic-rate tax) to a charity – and the charity can then claim back the tax from the Inland Revenue.

PART THREE

FAMILY MONEY

13 Buying a Home

How much can you afford? · Which type of property suits you? · The professionals – what they do and what they cost · Cutting costs · Shared ownership · Buying your council house

See also

- *Chapter 14, Mortgages*
- *Chapter 15, Selling Your Home*

For millions of people in this country, buying their own home will be the best investment they will ever make.

If you lived in London or the surrounding Home Counties during the last twenty years you'll have made and perhaps lost, your fortune out of your house. A typical three-storey terraced house in central London rose from around £20,000 in the middle of the seventies to nearer £250,000 in ten years. Over the next five years it probably dropped again to below £200,000.

More and more people are buying their own homes rather than renting them. Almost two thirds of the population are living in houses that they own outright or are in the process of buying, and that figure rises every year.

The main reason people buy rather than rent is because in the end you get something back for the thousands of pounds you put in. The monthly repayments on your mortgage are often not much higher than the rent on a similar house or flat. But instead of just lining your landlord's pocket, a mortgage will give you, after the twenty-five years or so you pay it, a home of your own. A piece of property that is yours to do what you like with. You can sell it and use the money to travel the world, continue to live in it, rent- (or mortgage-) free and pass it on to your children in your will.

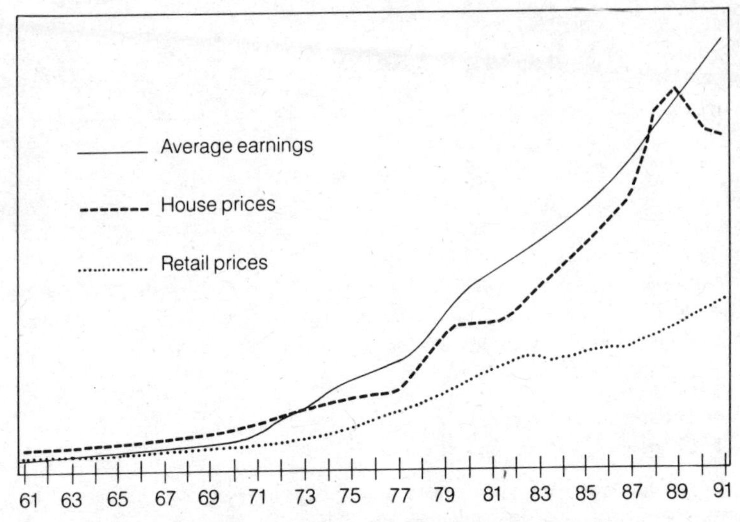

Indices of house prices, retail prices and average earnings, 4th quarter 1961 – 4th quarter 1991

YOUR HOME AS AN INVESTMENT

As an investment, in the short term, a home can be a non-starter. It won't pay you interest on the capital you sink into it; indeed your home will cost you money every year for running repairs and for doing it up.

The money you put into your home is there until you sell. You can't cash in a few bricks every time you are short of money or have an emergency you need ready cash for.

You have to pay your mortgage, month by month, year by year, in good times and in bad, or in the end the bank or building society will repossess.

But in the longer term, the capital growth on your home should see you through your old age. I say should because not everybody makes money on their house. In the late seventies and early eighties houses in some parts of the Midlands, the north-east and north-west of England and the central belt of Scotland proved hard to sell. And when the property market in the south-east turned from boom to bust in 1988 and 1989, some sellers found they were having to reduce the price of their houses by tens of thousands of pounds just to get potential buyers to come and have a look. Others gave up trying to sell at all, after 'For Sale' boards took root in their gardens.

If unemployment rises in your area, or your neighbourhood is blighted by the runway of a new airport or the building of a new power station, you could find that the value of your house goes down instead of up over the years. But generally, over the course of twenty to twenty-five years, the value of your home should rise at least as fast as inflation.

HOW MUCH CAN YOU AFFORD?

Be honest with yourself. There's no point in overextending your finances on the purchase price of the flat or house you choose and then having nothing left over to buy the furniture, pay the professionals' bills, foot the mortgage and maintain your house or flat.

The building societies used to have a strict lending policy when it came to mortgages. They'd offer two and a half times your salary plus one times the salary of any person you were buying with.

So if you earn £10,000 a year, and your partner £6,000 a year you'd be offered a mortgage of £31,000 (that's 2½ × £10,000 plus 1 × £6,000). That was the figure they calculated, on years of experience, that you would be able to repay out of the joint income.

Nowadays that has all gone by the board. Because so many more lenders have entered the game – the High Street banks, the insurance companies, the foreign banks – competition has hotted up. At one time, if you played your cards right (or wrong!), you could get up to six times your salary.

Tempting though that might be, you'd be unwise to take the money. The couple earning £16,000 a year might be offered a mortgage of £96,000 – which would mean monthly payments of £820 at interest rates of 10 per cent. If interest rates went up dramatically, taking monthly mortgage payments with them, that couple would be ruined. Unable to pay the mortgage and, as usual in times of high interest rates and economic recession, unable to sell their home, which will turn into an expensive albatross round their necks. Always remember that whatever you borrow you have to pay back.

If you are borrowing as a couple, a good rule of thumb is three times the salary of the higher earner and one times the salary of the lower earner. That should be your maximum.

Size of loan	Monthly repayments at 10%	12%	14%
£20,000	£150	£170	£190
£30,000	£224	£255	£286
£40,000	£309	£354	£400

So a man on £10,000 a year with a wife earning £6,000 should borrow no more than £36,000 (3 × £10,000 plus 1 × £6,000) and, at a mortgage rate of 10 per cent, that would mean monthly payments of around £270.

If you are borrowing on a single income, your net monthly payment should be less than a quarter of your total monthly income.

So a single person earning £10,000 a year should borrow no more than £27,800. £10,000 divided by twelve gives a monthly income of £833. A quarter of £833 gives £208 and that would allow you a mortgage of £27,800.

It is worth remembering, however, that you have to take your lifestyle into account as well. A married man with a couple of children and a wife who doesn't work will not be able to give up so much of his income towards the mortgage as a single man, because he has many more household and family expenses to come out of his salary every month. And if foreign holidays or an expensive hobby are a must for you, take those into account when working out the size of mortgage you can afford.

And remember that if interest rates go up and you have overestimated what you can afford, you could be finished financially. A rise of two percentage points on a £30,000 mortgage can add around £30 a month to your repayment bill.

CHOOSING YOUR PROPERTY

Once you know how much you have to spend, start looking around at properties at that price.

Deciding which area you look in will depend on your own circumstances. Only you will know whether you have to be near a school or shops, whether you want to buy in town or in the country, and which streets suit you for transport to work.

Sign on with the local estate agents and go and look at a lot of houses before you start to narrow the field.

New or old

New houses offer a higher level of mod. cons., beautifully fitted kitchens and bathrooms and a feeling of newness because no one has ever lived there before. Against that an old house will give character and probably more space, and is often much more conveniently situated for shops, schools and transport.

Watch out for developers offering houses and flats complete with fitted kitchens, bathrooms and often carpets and curtains as well. Although you will be getting all these fittings at a good price, you won't get your money back when you sell. The value of a second-hand cooker or dishwasher is very low indeed. And remember the premium you pay to move into a brand new house cannot be passed on to the next buyer either. So you may not make much of a profit if you move on within a year or two.

Flat or house

Choose a house if you want a garden and bit of privacy and don't mind being responsible yourself for keeping the house in order.

A flat comes with lower fuel bills, but it will also have annual service charges, which can rise steeply. A management company will be responsible for keeping the building and any grounds in good order so you won't have the hassle of finding a builder if the roof leaks, only the money to pay the bill. (And maybe a hassle with the management company instead!)

Freehold or leasehold

In England and Wales properties are currently sold in one of two forms – freehold or leasehold. With freehold you own everything, but with leasehold you lease the land and the building on it, that is your house or flat, for a set number of years. During that time you pay an annual ground rent – anything from £25 to several hundred pounds a year – to the owner of the freehold. At the end of the lease, the owner gets the building and land back.

Normally, houses are freehold and flats leasehold. And there is a sound reason for this. If you have a freehold, and your neighbour has a freehold, you have very little power over what the other person does with her building or garden (subject of course to planning consent and building regulations in the area). And you don't really need it. If she decides to knock down the internal wall between her kitchen and dining-room, that is up to her and will have no bearing on your house.

However, the position is reversed for leaseholders. Leases tend to be worded so that flat-holders do have some say in what other flat-holders in the building are doing. After all, if the chap on the ground floor decided to knock down the wall between his kitchen and dining-room and it turns out to be a load-bearing wall, it will have all sorts of consequences for the other residents.

So don't worry about not owning the freehold of your flat. Leases of longer than seventy or eighty years pose no problem but as the lease runs down it will affect the value of your property because the return to the owner looms ever nearer. By the time you get to less than thirty years the value of your home could actually fall year by year.

And never buy the lease of a flat from the owner without consulting your solicitor. Most building societies and banks are loath to lend money on freehold flats – for the reasons outlined above – and as most people need a mortgage to buy a property it could make it difficult for you when the time comes to sell.

New laws covering leaseholds in England and Wales will simplify this position. Once through Parliament, they will give flat owners with leases of more than twenty-one years the right to buy the lease either collectively as a block of flats or individually.

In Scotland, leasehold houses and flats are very rare. But many properties still have a 'feu duty' attached. It is rather like a ground rent in that it is a nominal amount paid annually, but in practice 'feu duties' are being phased out nowadays. Property owners can buy the 'feu' from the owner and, providing the price is right, you should have no worries about doing this.

Indeed before you sell a house or flat in Scotland you have to, by law, buy out the feu, so that, in a decade or so, feu duties will be gone for ever.

MAKING AN OFFER

Once you've got the feel of the housing market in the area in which you want to buy, check with your bank or building society that they will lend you the amount of money you are likely to need.

You can then go ahead and actually think about a specific property.

Once you've chosen a house or flat you like it is up to you to make an offer. I always think you are much better to deal directly with the estate agent, if one is involved, rather than the home owner.

You can offer the full asking price if that is what you are prepared to pay. But you won't do yourself any harm by offering less. Most house sellers expect potential buyers to offer a bit less than they ask, so they counteract that by asking more than they actually think their home is worth. In the recent years of the depressed housing market, many buyers are offering 10–20 per cent less than the asking price.

Haggling is not something that comes naturally to the British, and that is one reason why it is better to go through the estate agent when making your offer. It is much easier to offer him less, and get him to pass the message on to the house owner, than to ring the house owner directly and try to negotiate a cheaper price.

How much less you offer depends on the circumstances of the particular transaction. You may find you offer as much as you can possibly afford and give the owner a 'take it or leave it' ultimatum. You may be able to offer less than the asking price because there are several houses in the same road on the market, because you know the owner is a desperate seller or because the house has been on the market for a long time.

Remember you can always increase your offer if the initial price has been turned down, but you can't decrease it as easily if the owner accepts too readily. Whatever you offer, always make it 'subject to contract and survey'. That way you can renegotiate downwards if the survey throws up any ghastly problems such as unsound roof or dodgy drains.

In times of rapidly increasing house prices, you may have to offer more, of course, to avoid being gazumped. Gazumping is an appalling thing to happen to anyone. What happens is that after your offer has been accepted, and while you are getting mortgages set up and surveys arranged, another buyer makes a higher offer and the seller accepts it.

There is nothing, legally, wrong with this practice and some would argue that morally the seller has a duty to his family to accept the highest offer that is received. But if it happens to you, you will find it devastating – particularly if you have already paid for a survey on the property.

To avoid being gazumped, get moving fast. The further through that the deal has gone before a higher offer is made by someone else, the more likely the seller is to stick with the first offer. After all, it is a sale that is wanted, and a few hundred or even a few thousand pounds may not be too crucial.

Once your offer has been accepted you may be asked to pay a token deposit – this is normally returnable if the deal subsequently falls through. The £50 or £100 should be lodged with the estate agent, or solicitor, never with the house owner, and you should always get a receipt.

For the house seller, the problem now is inverted gazumping. Many is the hopeful seller that gets right to the point of exchanging contracts before being asked to dock anything from £500 to £5,000 off the asking price. It is usually a take-it-or-leave-it offer, with the buyer threatening to move on if the new price is not acceptable.

BRINGING IN THE PROFESSIONALS

Your next step is to bring in the professionals – a solicitor, a surveyor and of course a bank, building society or mortgage lender to provide the mortgage.

Most people don't have these sorts of professionals on tap so it is often difficult to know where to find one. Start by asking around among your friends to see if they have a good solicitor, who perhaps did the buying of their house for them. Failing that, get out your phone book and pick a few names at random and phone up asking what they charge to do the work involved in buying a house. Using the solicitor that is acting for the mortgage lender can cut your legal bills.

The solicitor or licensed conveyancer

Although you can do all the legal work (known as conveyancing) yourself, most house buyers use a solicitor or licensed conveyancer. It protects them from making what could be very expensive mistakes.

In the past all conveyancing was done by solicitors, but a change in the law has opened up this area to licensed conveyancers.

Charges vary from expert to expert but the more expensive the house and the more work he has to do, the more your solicitor or conveyancer will charge. I would not advise you to do your own conveyancing. It is a complicated and time-consuming business and, unfortunately, you won't know whether or not you've done it correctly until it comes to selling the property. Then if you did make mistakes on buying you could find it takes a lot of money to put them right.

It is also advisable to find out what the solicitor is likely to charge, before you commit yourself. Phone around several and ask for a guide to the likely cost and that way you could save yourself several hundred pounds.

> **Tip**
> A recent survey shows that written estimates often save you money because solicitors keep to their original calculations. So get your estimate in writing.

There are five main tasks your solicitor will do for you:

1 Search

He'll send a list of questions to your local council to check on any future developments in your area, such as a motorway at the end of your drive or a housing development on nearby open fields.

2 Land registration

Over three quarters of the homes in this country are already registered with their local District Land Registry (and it is mainly rural ones which are not). That means that the seller's title to the house has been checked, and it is virtually guaranteed, by the government. If the house you are buying is registered it is a simple matter for the solicitor to check the entry. If it has not been registered yet your solicitor will have to do a bit more checking and register the house.

3 Stamp duty

If the house or flat you are buying costs more than £30,000 you'll have to pay a tax of 1 per cent, known as stamp duty, on the full price of the property to the government. For some time in 1991 and 1992 stamp duty was lifted on properties under £250,000 to try and breathe some life into the housing market.

4 Exchange of contracts

Once the actual house buying/selling deal is almost complete and the surveys and search are completed and the mortgage in place, you and the seller will exchange contracts. This binds you, legally, to buying the property and once you've signed the contract and it has been exchanged you cannot back out. You will also have to pay a full deposit, usually 10 per cent of the agreed price of the property.

5 Completion

This usually takes place a month after the exchange of contracts and formalizes the sale. It all happens at your solicitor's office. You don't have to be present when he exchanges documents and hands over a cheque for the balance of the cost of the house. In return, he'll get the keys. You can then move in.

The Scottish system

The system in Scotland is slightly different – and many would argue much better than that in England and Wales, because it prevents gazumping.

Properties in Scotland are sold through a bidding system, with no bidder knowing what anyone else has offered. The offers are all opened at the same time and the highest bid wins the property. You could be lucky and get your new home for a few pounds more than the next person was willing to pay, you might find you've bid a few thousand pounds more. Either way, you are committed.

Once your lawyer has sent a letter of offer (your bid) and received a letter of acceptance (it is known as an exchange of missives), you and the house seller are committed to the deal.

It is essential therefore that you set up your mortgage and arrange the surveys before you make an offer on a property. This can become expensive if you bid for, and lose, several properties.

Surveyor

Once you have made an offer, and engaged a solicitor, you'll need to make sure that your potential home is sound.

To do this you'll need a surveyor.

It is often tempting, in the expensive business of house buying, to skimp on your survey, since it can cost upwards of £200. But you could pay dearly for this saving. There is no obligation whatsoever for the seller to point out any of the defects of the house. And if you don't spot unkeyed plaster or a corroding water cistern you could end up with bills of thousands of pounds that you didn't expect once you've moved in.

Choosing a surveyor

Always, always choose a qualified surveyor. The letters FRICS or ARICS after his, or her, name will show that he or she is a member of the Royal Institution of Chartered Surveyors. If you want to check up that the letters are genuine phone the society on 071 222 7000 for England and Wales, 031 225 7078 for Scotland. Alternatively the surveyor may be ISVA, a member of the Incorporated Society of Valuers and Auctioneers.

Types of survey

There are three types of survey that you can choose from. As a general rule, the more you pay, the more information you get. It is up to you which you go for.

1 Valuation

The building society or bank that offers to lend you the money to buy the house will ask for a valuation of the property. They insist on it, you have to pay for it and it ensures that the property you have chosen is adequate security for their loan. No more, no less. Many societies and banks will send you a copy of this report; if they don't, you should ask for it. After all you've paid for it.

On a £75,000 house it would cost around £125.

2 House buyer's report

This is a halfway house of a survey which will show up any major defects. If you use the same surveyor who is doing the valuation for the lender he'll do your house buyer's report at the same time, and that should cut costs a bit. The report will include details on the state of repair and condition of the property. It is not totally comprehensive but it may suit you if the house is not too old and has been well maintained.

On a £75,000 house it would cost around £250.

3 Full structural survey

This is the type of survey you should have done if you can afford it. Again you might be able to trim costs if you use the lender's surveyor. In this case the surveyor will check for damp, for any structural defects, he'll look at the brickwork, the woodwork, the drains, the roof, the rendering, even the boundary walls or fencing – in fact, give the house a thorough check.

And if he does find any major faults that will require a lot of money to put right, you may be able to use his report to negotiate a reduction in the asking price of the house.

On a £75,000 house it would cost around £375.

But remember, a surveyor is not infallible. Have a good look at the property yourself to see if the walls are bulging, the window frames rotten, the ceilings showing the tell-tale water marks of a leaking roof or the corners of the walls showing the black spot signs of condensation. If you've spotted any of these defects, and there's no mention of it on the structural survey, give your surveyor a ring to ask him why.

BUILDING SOCIETY, BANK OR MORTGAGE LENDER

When you've made your successful offer, tell the building society, bank or mortgage lender from whom you are getting your loan. You'll be asked to fill in an official mortgage form at this stage – and now the costs of getting the mortgage will begin to mount.

- You may have to pay an arrangement fee of up to £100, sometimes more, particularly if you are dealing with a bank.
- There will be the cost of the valuation that the lender will insist on.
- You will have to pay the legal costs of the building society, bank or mortgage lender.
- If you are asking for a mortgage worth more than 75 per cent of the property you will probably have to pay an indemnity fee. An indemnity fee is what you have to pay to prevent your bank, building society or mortgage company having to shoulder the whole risk should you default on your payments. Most lenders prefer to limit their risk to less than 75 per cent of the value of the property. If you want a larger loan than that they'll reinsure the rest with an insurance company. That way the lender reduces its risk, but you have to pay for it. On a £75,000 house that could cost around £1,000.

HIDDEN COSTS

You may think that once you've got your mortgage lined up at an affordable level, paid the professionals for their part in your purchase, and used up all your savings in the process, there can be nothing left for you to pay for.

Wrong. There's still a multitude of bills to be paid.

There will be the removal costs. Even if you are a first-time buyer you may need to hire a van or pay for a removal firm to take you from your rented flat to your new home, or you may be buying second-hand furniture which will need moving.

If you are actually moving house, the costs will be much higher. How much higher will depend on the amount of furniture you have, whether or not you pack it yourself, and how far you are travelling. It is certainly worth getting several quotes before committing yourself but if you want to check out the firm you choose, give the British Association of Removers a ring on 081 861 3331.

Then there are the services: gas, electricity, water and phone. If they're still connected make sure that someone reads the meter on the day of the move. If they've been cut off, reconnection charges will be made.

Once you've moved into your new home, the bills will continue to pile up. You'll have house insurance, both for the building (since you exchanged contracts) and the contents, water rates and local authority charges, repair bills, and bills for redecorating. You may find fuel bills are higher than you are used to and if you have a garden you'll find out just how expensive that is to keep neat and tidy. So it is very important not to overextend yourself with your mortgage or you'll come seriously unstuck when these bills start to mount.

How to cut costs sensibly

If you haven't bought your first home yet the thought of all those bills might be putting you off. Don't panic. There are several ways to cut costs sensibly and legally.

1 When you have negotiated a price with the seller that is acceptable to you both make sure that it is split into the price of the house and the cost of the fixtures and fittings – any carpets, curtains, kitchen and bathroom goods and so on that you are buying from them. That is

particularly important if the house or flat you are buying costs just over the £30,000 mark because it will save you a few hundred pounds in stamp duty. For example, if you pay £30,500 for your flat you'll have to pay £305 stamp duty. But if you pay £29,900 for the flat, and £600 for the carpets, curtains and furnishings you will have no stamp duty to pay at all. Don't try paying £29,900 for a flat and a ridiculous £15,100 for the furnishings of a £45,000 flat, though. The Inland Revenue won't wear it.

2 Unless it is absolutely essential, don't rush the sale through. If your lawyer has to take expensive short-cuts, like using bikers to deliver contracts instead of the Royal Mail, or work overtime himself, he'll charge you all the extras.

3 If you find out about a house sale in any way other than through the estate agent, and you don't involve the agent, you might be able to persuade the seller to pass on half the estate agent's commission that he won't now have to pay. Many sellers, though, prefer you to use the estate agent.

4 If the house needs a lot of improving, check with your local authority before you start any work to see if there is a chance of getting an improvement grant. Some are yours by right, others depend on the amount of cash the local authority has on offer – usually not very much. If your house comes within the rules for qualifying, it is always worth applying. Ask for more information at your local town hall.

5 If you are a first-time buyer, work out in advance how much your mortgage is going to cost you. Deduct the rent you currently pay, and while you are looking around or negotiating to buy, deposit the balance every month into a savings account. That way you'll get used to living on a reduced income, and know, before you finally commit yourself, that you can cope. You'll also be building up a tidy nest egg which will be very welcome when the bills come in.

6 Don't let your bank, building society or mortgage lender push you into buying house contents insurance or buildings insurance from a company they specifically recommend. It may not be the cheapest for you. They always have a selection of companies whom they are happy for you to deal with and it will be worth your while shopping around for the best deal. Your bank or building society may, however, charge an administration fee if you arrange your own insurance, so check that in advance.

SHARED OWNERSHIP

If you can't afford to buy a home of your own, but still want to get a foot on the housing ladder, shared ownership may be the answer. It is not the same as a shared mortgage, where you and a friend or relative club together to buy a place. Shared ownership means you only buy part of a house or flat, and pay rent for the rest, and you will deal with a housing association.

Initially you can buy as little as 25 per cent or as much as 90 per cent of your new home, but you can increase this stake right up to 100 per cent when your finances allow it.

It doesn't mean you are committed to living in the house until you own the whole 100 per cent. You can sell whenever you want. You can either sell your part of the property on to someone else wanting shared ownership or you can sell the property outright to the new owner while simultaneously buying the property outright from the housing association.

Shared ownership started in the mid-seventies in Birmingham and it is possible to find schemes in most areas now though queues may be long.

Housing associations, new towns and local authorities tend to offer shared ownership housing projects but you can get further details of schemes and a helpful booklet from the Housing Corporation, 149 Tottenham Court Road, London W1P 0BN (tel. 071 387 9466).

Shared ownership is invaluable for first-time buyers who couldn't otherwise afford to buy at all.

BUYING YOUR OWN COUNCIL HOUSE

Over a million people in this country have already bought their council houses – most of them taking advantage of the hefty discounts that are available.

Providing you continue to live in your ex-council house for three years after you have bought you can get a discount of up to 60 or 70 per cent on the market value up to a maximum of £50,000. The size of the discount depends on whether you live in a flat or maisonette or in a house, and on how long you have been a council tenant.

A council tenant who has lived in a flat or maisonette for thirty years does best, with a 70 per cent reduction on the price she will have to pay, but even someone who has only been a council tenant for two years and

Qualifying period (years)	*Discount (%)*	
	Houses	*Flats/maisonettes*
2	32	44
5	35	50
10	40	60
15	45	70
20	50	70
25	55	70
30	60	70
over 30	60	70

lives in a council house will get almost a third off the buying price. You don't have to have been living in the same home, or even been a tenant with the same council for the length of time you are using to qualify for a discount.

How to buy

To buy your council home you'll have to take a number of clearly defined steps, most of which involve filling in oddly named and numbered forms. First you must fill in form RTB1. Your town hall must supply you with a copy but if you have any difficulty getting one, write to the Department of the Environment (if you live in England), to the Welsh Office, or the Scottish Office and they will send you one.

Once you've filled in the form (and I'd recommend keeping a copy for yourself), send it back to the town hall and you should receive, within four to eight weeks, a copy of form RTB2 confirming your right to buy.

Over the next eight to twelve weeks you'll get a Section 125 Notice, telling you how much your home is being offered to you at and what conditions are attached to the sale.

The price will be the market value of the property, less any discount you are entitled to.

If you think it is overpriced you can appeal to the District Valuer, but if he finds that, in fact, the house has been undervalued you will have to accept his higher price for the property if you go ahead and buy. So don't automatically appeal.

Although the council must, by law, tell you if there are any structural defects on the property, you should get your own survey done in case there are any gremlins you don't know about.

Then you just complete in the same way that you would if you were buying a house from an individual.

What if you can't afford your house?

You can put off buying your council house for up to three years after you apply to buy and the conditions won't change in that time. Or you can go for shared ownership (see above) so that you buy part of the house and rent part.

Selling your council house

You can sell your ex-council house any time you like but if you move within three years you'll have to pay back part of the discount.

sell within	1st *year*	*repay* 100% *of discount*
	2nd	66.6%
	3rd	33.3%

Should you buy your council home?

Whether or not you decide to buy will be a personal decision based on more than purely financial considerations. You might want to have something to pass on to your children, or just not like paying rent year after year and getting nothing back for your investment. Or you may want to see the value of your property rise, which it almost definitely will do if you qualify for a particularly large discount.

But do your money sums as well. Owning your house can prove more expensive than renting if you have a large mortgage.

Use the table on the next page to work out your monthly costs.

You should be able to work out, from comparing the totals, whether or not you can afford to buy your home.

Getting a mortgage

You will be able to get a mortgage just as easily as anyone else, providing you can prove that you have the money to keep up with the monthly repayments. For full details read the next chapter.

Buying a house	£	*Renting a house*	£
Mortgage		Net rent	
Council charge		Council charge	
Water rates		Water rates	
Insurance			
Repairs			
Total	£	Total	£

Buying a flat	£	*Renting a flat*	£
Mortgage		Net rent	
Council charge		Council charge	
Water rates		Water rates	
Service charge			
Insurance			
Repairs			
Total	£	Total	£

Other tenants who have the right to buy

You have the right to buy the house, flat or maisonette you live in if you rent it from one of the landlords listed below; if you've been living in it or in a property rented from another 'right-to-buy' landlord for at least two years; and if the house or flat is your main home.

However, there are a whole list of exceptions. You can't buy a home that has been specially converted for the old, or the mentally or physically handicapped, or a house that is being used as temporary housing pending redevelopment or a tied house. Squatters are also excluded.

Right-to-buy landlords in England and Wales

- a district council
- a county council
- a London borough council
- the Common Council of the City of London
- the Council of the Isles of Scilly
- the Inner London Education Authority
- a metropolitan county police authority
- the Northumbria Police Authority
- a metropolitan county fire and civil defence authority

- the London Fire and Civil Defence Authority
- a metropolitan county passenger transport authority
- the London Waste Regulation Authority
- the Merseyside and Greater Manchester Waste Disposal Authorities
- the London Residuary Body
- a metropolitan county residuary body
- a new town or urban development corporation
- the Commission for the New Towns
- the Development Board for Rural Wales
- a housing association that is registered with the Housing Corporation and is not a charity, an association not getting public subsidy or a cooperative association
- the Housing Corporation
- the West London, North London, East London and Western Riverside Waste Disposal Authorities.

If you are thinking of buying your rented home it is worth getting a copy of a leaflet called 'Your Right to Buy Your Home'. It is available from the Department of the Environment, the Welsh Office or the Scottish Office.

STEP-BY-STEP CHECKLIST TO BUYING A HOUSE

1. Work out how much you can afford from mortgage and savings.
2. Deduct cost of legal fees, valuations, stamp duty, etc.
3. Find house.
4. Make an offer.
5. Pay token deposit.
6. Make formal application for mortgage.
7. Instruct solicitor.
8. Get house surveyed.
9. If necessary, renegotiate price or look for alternative property.
10. Check that solicitor has found no hitches.
11. Exchange contracts.
12. Check with gas, electricity and water authorities and British Telecom that meters will be read just before you move in.
13. Complete.
14. Move in.
15. Open bottle of champagne.

Extra costs

- stamp duty
- solicitor's fees including fee for purchase, Land Registry, searches
- mortgage-related costs
- survey and valuation fees
- building and contents insurance
- removal costs
- reconnection fees.

14 Mortgages

The types of mortgage available · How to choose the best one for you · What to do with your loan when you move home · Coping if you can't pay the mortgage

See also

- *Chapter 6, Life Insurance*
- *Chapter 13, Buying a Home*
- *Chapter 15, Selling Your Home*

Buying your own home and getting a mortgage go hand in hand nowadays. Very few people can afford to pay cash for their first house or flat.

Over 9 million people have a mortgage on their home and among them they've borrowed a staggering £320 billion.

A local building society used to be the first stop when anyone wanted a mortgage. There they pleaded with the manager to be allowed to borrow the money and usually had to prove they had saved regularly for at least two years. That sort of attitude has probably now gone for ever. Increased competition in the mortgage market from banks, insurance companies, mortgage companies and pension funds, coupled to a downturn in the housing market, has turned the tables on the building societies. Now you'll be welcomed with open arms by a lot of lenders, and offered a wide array of deals.

Your home loan can give you a helping hand to buy the house of your dreams or become a nightmare debt you are saddled with for twenty-five years, so it is vital that you choose carefully and don't overborrow.

WHAT IS A MORTGAGE?

A mortgage is just a long-term loan (with tax relief), which you agree to pay back monthly over an agreed length of time, usually twenty-five years. You will have to pay interest on the money, and put your home up as collateral. That means if you don't or can't pay back the loan the institution that lent you the money will, as a last resort, take over and sell your property to get their money back.

TYPES OF MORTGAGE

Although there are about a dozen different types of mortgage on the market don't be put off – they are all variations on two basic themes: the repayment mortgage and the endowment mortgage.

Repayment. With a repayment mortgage you borrow your sum of money and pay it back, a little at a time, with the interest over the length of the loan. So every month you pay back some of the capital and some of the interest.

Endowment. With an endowment mortgage you borrow the sum of money and pay the interest only over the length of the loan. At the same time you take out an insurance policy into which you pay enough money to build, with its investment income, a large enough lump sum to pay off the mortgage at the end of the agreed term. So every month you make an interest payment and an insurance payment.

There are various types of repayment and endowment loans, designed by the banks and building societies to suit the various types of borrowers that they have. One of them will suit you best.

Repayment

Constant net repayment. In the long run this is the cheaper of the two basic types of repayment mortgage. Monthly repayments are calculated so that the borrower pays exactly the same every month for the life of the mortgage, assuming that interest rates don't change. (In fact they will go up and down but the principle still stands.) So your mortgage will be as expensive in the early years – when you can probably least afford it – as in the later years.

Gross profile. This type of mortgage is cheaper in the earlier years but in the long run is likely to be more expensive than the constant net repayment loan. The borrower has to pay back more interest and less capital at the beginning than under the constant net repayment method, and because she gets tax relief on the interest (see MIRAS, p. 162) her repayments will be lower. But the capital element in the loan is not going down as fast because it is mainly interest that is being paid in the early years, so in later years the repayments will remain higher because the interest on the larger capital sum will still be quite high.

Gross profile loans are helpful for first-time buyers, who are often stretched financially in the early years of a mortgage but hope to be better off in the coming years when their salary goes up. But watch out if you are buying as a couple and intend having children in the foreseeable future. Your repayments will probably be going up just as one income is lost if the wife leaves work to have a baby.

Endowment

With-profit endowment. Premiums are pitched at a high level so that at the end of the term of the mortgage the lump sum will not only pay off the mortgage but will also leave the borrower with a substantial nest-egg. Nice, but the most expensive of all the schemes and not much used nowadays. If you want to save through a with-profits endowment plan, do so, but I don't think it is a good idea to link it to your mortgage.

Low-cost endowment. Has the advantage of much smaller monthly payments than its with-profits cousin, but there is no guarantee what size the lump sum will be at the end of the twenty-five years (or whatever length the mortgage runs). The lump sum could actually be smaller than the size of the mortgage and the borrower would be left to make up the difference. This would be unlikely, but not impossible, and most companies offering these plans expect the lump sum to be bigger than the mortgage, leaving a small surplus.

Non-profit, or guaranteed, endowment. This guarantees that the mortgage will be paid off at the end of the term, but gives no surplus to the borrower. It is poor value for money and a last resort.

Low-start, low-cost endowment. As the name suggests borrowers pay less into this policy at the start of the mortgage. After five years or so of steadily rising premiums, repayments level off at a slightly higher rate than those of the ordinary low-cost endowment. Good for first-time buyers and others having difficulty affording their mortgage but the same reservations apply as those on the gross-profile repayment loans. You could find repayments going up just as you lose one income if you decide to start a family and your wife gives up work.

Alternatively you could borrow an extra £5,000 to £10,000 and use that to subsidize monthly payments for the first three years. Then payments would revert to the normal, larger, size and the lump sum would have gone. Be wary of this option.

Pension mortgage

There is a huge tax advantage to having a pension mortgage and they are available to the self-employed with their own pension schemes, and to employees who have a personal pension. Like an endowment mortgage you still pay monthly interest to the building society or bank from whom you borrowed the lump sum, but alongside that you pay into a pension plan which will pay off your mortgage on retirement and leave you with a pension. The tax advantage is that all contributions are tax-free, and you can pay in up to 17.5 per cent of your annual earnings – much more if you are over fifty. Your money should build up faster in this type of fund than in an ordinary endowment fund because pension funds don't pay capital or income tax on their investments.

The disadvantage is that you will get a much lower pension in retirement if a large part of your lump sum is used to pay off your home loan.

Unit-linked endowment mortgage

These mortgages are for people wanting to take a bit more of a risk with their premiums. The money that you pay into your endowment policy is invested in unit trusts and because the price of units goes down as well as up your home loan is linked directly to a riskier proposition. Over the course of a twenty-five year mortgage you would be unlikely to lose out on your premiums but the investments might not rise fast enough to cover the size of lump sum needed to pay off your mortgage. You could be asked to step up your monthly contributions part-way through the term of the loan. However, the potential for making a much larger lump sum is also there and if the fund managers do their job well and share prices are rising, you would be able to pay off your mortgage much earlier than you expected. None the less, this type of mortgage is for people who like to live a little dangerously.

Fixed-rate mortgage

This loan offers borrowers a chance to fix the rate of interest they pay for a certain number of years. No matter what happens to interest rates in general, your mortgage rate will stay at whatever rate was fixed when you took it out. So:

- if interest rates go up you win because you are paying less than everyone else
- if they go down you lose because you are paying more, but whatever happens to rates at least you know exactly what your monthly payments will be over the years of the fixed rate.

The rate is not usually fixed for the whole twenty-five years of your mortgage; usually the term will be one to three years, then a new fixed rate will be offered to you or you can switch to the normal variable mortgage rate. The great advantage is that you know exactly how much your mortgage will cost you every month.

PEP mortgage

A revamping of Personal Equity Plans – allowing investors to pay in up to £6,000 a year – has made PEP mortgages a buzz word. They work like endowments in that you borrow the money to buy the house and pay interest only on it for the life of the loan (usually twenty-five years). You then build up a fund to pay off the loan at the end of the term.

The great advantage of a PEP over an endowment is the tax one. PEPs are much more tax-advantageous and more flexible. You can take money out throughout the life of the mortgage to pay part of it off without penalty. That way you would reduce the amount you are paying in interest every month. Always a welcome move!

You are, however, limited to the amount you can pay into a PEP. Nowadays your entire PEP entitlement – that is, £6,000 – can go into a unit or investment trust. And that would be the most cost-efficient and risk-free way of building up your endowment-type fund. But it should still be enough, over twenty-five years, to cover a £200,000 mortgage.

Life insurance and mortgage protection

All endowment policies automatically carry life insurance. If the policyholder dies, the mortgage will be paid off in full and any extra money will be paid into the policyholder's estate.

Repayment mortgages carry no such insurance. So most people who take out a repayment mortgage should also get a mortgage protection plan. It is in fact a decreasing term insurance policy, and for a woman in her thirties will cost only a few extra pounds a month. Because of the Aids scare, it is more expensive for a man, particularly a single man in his thirties (see Chapter 6, Life Insurance).

In the event of the death of the policyholder the mortgage will be paid off but there will be nothing extra.

Mortgage flow chart

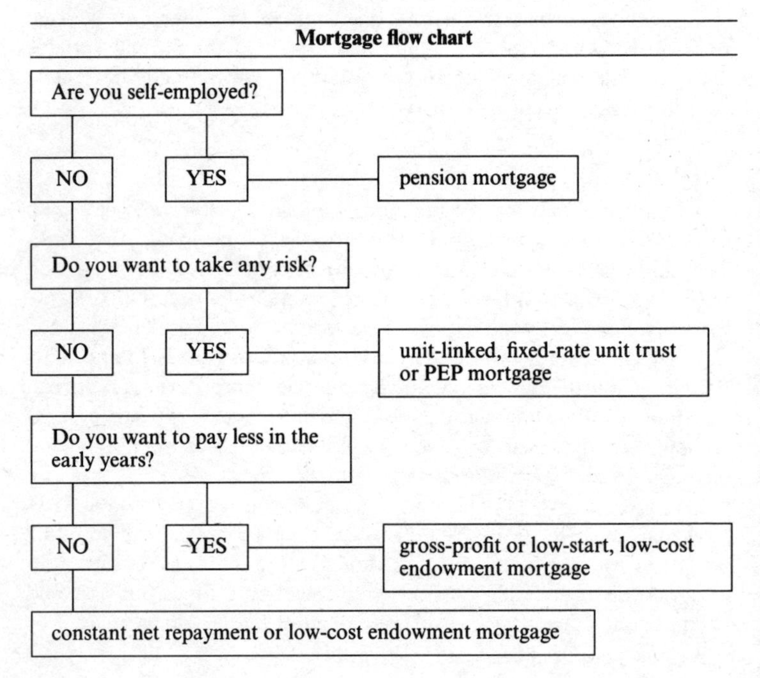

WHICH TYPE OF MORTGAGE IS BEST FOR YOU?

Repayment v. endowment

Whether you choose a repayment mortgage or an endowment mortgage depends on your own financial circumstances. Because of the way mortgages are calculated, when the mortgage rate is less than 10.5 per cent an endowment is better value; once it gets over 10.5 per cent then a repayment is cheaper. During the life of your mortgage the rate will likely be both above and below 10.5 per cent so cash-wise they will both work out much the same.

Go for an *endowment* if:

- you want the chance to build up a surplus lump sum.

Go for *repayment* if:

- you want flexibility. If interest rates rise sharply you can always extend the term of the loan so that you continue to pay the same monthly repayments – that option is not always open to you with an endowment mortgage. If you have difficulty meeting the payments, the building society or bank may allow you to pay interest only for a short while. You couldn't do this with an endowment because if you stop making contributions to the insurance policy and surrender it, you may get back less than you put in.
- you are getting on in years. As you get older, taking out a new endowment policy becomes increasingly expensive as the odds shorten on the number of years you are likely to survive.

Warning

Anyone who sells you an endowment mortgage will get a nice commission on the deal: around £600 for arranging a £30,000 endowment mortgage. Don't believe their projections for the size of the lump sum you might get at the end, they are forecasts, not fact. And don't forget that inflation will eat into that sum. £20,000 may seem like a lot when you take out the endowment, but in twenty-five years' time you may find that it will only cover a two-week holiday on the Costa Brava, who knows. Buy an endowment only if you want one, not because someone is selling it to you.

Where to go for a mortgage

You can get a mortgage from a building society, a bank, an insurance company, a mortgage company or a local authority.

The difficulty is not getting the money but getting the best value for money.

Over the past three or four years the 'mortgage rate' has disappeared. In its place has come a mortgage level. The lenders set their own mortgage rate and generally they are within 1 to 1½ per cent of each other. But it does mean that at various times a bank or building society could be undercutting someone else's rate by as much as 1½ per cent. And that it is a lot of money on a £20,000 to £30,000 mortgage, particularly if it is not your lender who is doing the undercutting.

It is impossible to guess who is going to have the lowest rate in the years to come. The only thing you can know is who has got the cheapest rate now, and that is the best place to start when you need a mortgage.

But, as in all things financial, nothing is quite that simple. It is important to find out in advance about additional charges.

Some banks charge an arrangement fee up to £100 for giving you the loan, other lenders insist on three months' notice, or an extra three months' worth of mortgage payments if you want to redeem the loan (which you will want to do when you move house).

Some lenders still charge more on large loans, others offer a discount; still others penalize you if you want to borrow small amounts of money, so check that their mortgage rate applies to all customers, or find out what the small print says.

Start with the bank and/or building society with which you have an account but look at their competitors, and the mortgage companies, too. With luck, at the time you are looking for a mortgage, one of the lenders will have a special deal running. They might be offering a reduced mortgage rate in the first year, free life cover, a waiving of legal fees, indeed anything that will make your first year's bills lower.

These schemes are often worth taking up – providing the lender is offering you a mortgage that suits you anyway and you are not being forced to buy something you don't want.

If you choose an endowment mortgage, you will also need an endowment plan to go with the lump sum you borrow from a bank or building society. Get several quotes from the large insurance companies and go with the one which offers you the best value for money. But remember the sizes of the lump sum being forecast at the end of the term are no more than a projection.

Local authorities also provide loans for house buying – indeed they have a duty to do it if you are buying a council house. But their rates are often higher than you would get from a bank or building society.

HOW MUCH TO BORROW

Don't be sucked into overborrowing just because someone will lend you the money. A mortgage is not a short-term loan that you will somehow cope with, like a large credit card bill; it is a long-term debt that could sink you financially. So be realistic when working out what you can afford. Your bank or building society will probably limit you anyway to what they think you will be able to pay back but I'd advise you to stick within the following limits:

- try not to borrow more than three times the main income plus one times any secondary income; that means that if a couple are buying together and the man earns £10,000 a year and the woman £7,000 a year (or vice versa) you should limit your mortgage to £37,000
- ensure that your net monthly payments are less than a quarter of your total monthly income – the couple on £17,000 a year should keep their payments below £350 a month.

Watch out if you are following these ratios slavishly, however – they are a bit old hat now. You also have to write your lifestyle into the equation. In its simplest form, a married couple with children who live solely on the husband's income cannot afford to have the same percentage of that salary going towards the mortgage as a single man could. The married man's salary has to foot a lot more bills.

If you value your holidays or a hobby, remember to put that money aside before deciding how much you can afford to repay every month. And if you have a dicey job that could fold up under you, don't over-mortgage. That way you should still be able to afford to eat, once you own your home.

And always try to pay at least part of the cost of the home yourself. Most lenders will want you to contribute at least 5 per cent and possibly up to 10 per cent of the purchase price but if you are a first-time buyer with no profit from another property this might be difficult. If you can't put anything into your home yourself you might be forced to take on quite an expensive top-up mortgage from an insurance company to make up the difference. First-time buyers can usually get 95–100 per cent mortgages, however.

So if you are planning to buy a flat or house in the near future, start saving now.

MIRAS: tax relief on your mortgage

If you borrow money to buy your own home, you will qualify for tax relief on the interest payments up to a ceiling of a £30,000 loan. The tax is deducted from your monthly payments before you pay them and the system is known as MIRAS – Mortgage Interest Relief at Source. You no longer get tax relief at your top rate of tax.

No matter what size the mortgage, tax relief stops at £30,000 per person, or per married couple. Unmarried couples or friends sharing a mortgage can no longer get tax relief of £30,000 each, although existing deals still stand.

What the lender will want to know about you

Before lending you money your bank or building society will want to know that you are a sound financial bet.

Information will be needed on:

- your earnings
- your age – particularly if you want an endowment mortgage or a mortgage protection plan
- your health
- your financial record – have you a lot of debts, a good savings record
- other outstanding loans.

What they are trying to work out is your ability to pay your mortgage every month. Although they could, in the end, sell your house to get their money back no lender wants to do this if they can avoid it. So, although they have the collateral of your home, they will also ensure that you can afford the loan.

Few banks or building societies nowadays will refuse you a loan because you are a woman or because you are too old. Pensioners can take out ordinary mortgages like the rest of us (though they would probably choose a ten-year term rather than a twenty-five-year one), providing they can prove that their pension is large enough to cover the monthly payments.

MOVING HOME

The mortgage

When you move house one of the last things you think about is what to do about the mortgage. Think about it first or it could cost you quite a lot of money.

First of all find out if your lender charges you for 'early redemption'. That means for stopping the mortgage before the end of the term. Some lenders ask you for three months' notice or they charge an extra three months of repayments if you cash in your mortgage in the first few years. If yours is one of those, make sure you give the notice in plenty of time.

Repayment mortgage

With a repayment mortgage the simplest thing to do is pay it off when you sell your home and take out a new mortgage on your new home.

Endowment mortgage

With an endowment mortgage you have a choice. You can cancel your endowment policy, make it 'paid up' or transfer it to the new property and take out a new endowment policy for the difference if you need a bigger mortgage. (See Chapter 6 for explanations of terms.) If you surrender an endowment policy in the early years you may not get back even the contributions you have paid in, let alone any growth on that money.

I would advise you to transfer your endowment policy, and remember you will be a few years older when you take out any additional policy, so the terms will not be so generous.

Moving your mortgage

If you are not satisfied with the deal you have, you could move your mortgage. This sounds easy in theory. In practice it is not so simple.

You can move from a repayment mortgage with one lender to a repayment mortgage with another lender. Or you can move from a repayment to an endowment mortgage. Changing from an endowment to a repayment is not such a good idea because if you cash in the endowment policy you will almost certainly lose on the deal. Any top-up you have had for home improvements will lose its tax relief – only the original loan will qualify.

And, unless there is a special deal available from the lender to whom you are moving and they are willing to pay all your expenses, you'll find that the move costs you a couple of hundred pounds. There will be legal fees for the transfer of the mortgage and invariably the new lender will ask for a valuation, if not a full survey of your home, and you will have to pay for that too.

So unless you are sure that the new mortgage is going to remain much cheaper you are probably better off staying with the lender you already have.

Bridging loans

If you buy your new house before selling the one you are in you will need a bridging loan to tide you over. In fact, they are normal practice in Scotland. Bridging loans can be a good thing, for example if you have exchanged contracts with the buyer of your home but want to move into the house you have bought a few days or weeks earlier. A bridging loan will enable you to do this and you won't be crippled by long-term debt because you know when the money is coming through from the sale of your old house.

What you should not do, in England and Wales, is use a bridging loan to buy a new property before you have sold, and exchanged, on the old one. Bridging loans are expensive – interest is usually at around the overdraft rate – and if you don't know how long the loan is for you could find the payments very expensive indeed, as you will still be paying the mortgage on your old property as well.

If you do take out a bridging loan, remember to tell the Inland Revenue about it because you can claim tax relief on the bridging loan up to another £30,000 – that is in addition to the £30,000 tax relief you already get on your mortgage. And if you have difficulty selling your house, you can claim this tax relief for one year at least, often for two years.

WHAT HAPPENS WHEN INTEREST RATES CHANGE?

If interest rates move up or down, your bank or building society will notify you of the change and one of two things will happen.

1 Your monthly payment will change. If interest rates go up you'll have to pay more per month. If they come down, you will pay less.
2 Your monthly payment will stay the same. Some lenders operate a policy of only changing the amount you pay once a year. So if interest rates change you will continue to pay the same and the change will be incorporated in the following year's payments.

> **Tip**
> If you have an annual change, you can move back. This can be a help when interest rates are coming down.

In the long run you will pay the same amount of money either way.

WHAT HAPPENS IF YOU CAN'T PAY THE MORTGAGE?

If interest rates rise sharply, or you overspend on your other finances or you have a family trauma, you could find yourself unable to pay the mortgage. There are several steps you must take quickly to avoid serious trouble.

The first step is to write to or go to see the building society or bank manager, or your mortgage company. They are used to this happening and will try to help you. It is easier in the long run for them if they can help you through these bad times rather than have to sell your house to get their cash back.

If you just stop paying the mortgage the manager will summon you and you won't get such a positive reception when you turn up. And don't think nobody will spot it if you don't pay every month. The computers are set to notice if anyone misses more than one or two payments.

What happens next depends on whether your financial problems are short term or long term. For example, if the breadwinner has lost his or her job, are they likely to be unemployed for a month or two while they find other employment or on the dole until retirement age?

Short-term problems

This is the easiest financial crisis to cope with. Temporary unemployment, perhaps an illness in the family or a mad spending spree can hit your finances hard. If you can go to your building society or bank manager and tell him what has happened and why you can't pay the mortgage for the next month or two, he may agree to roll over the payments for up to three months. That means that you won't have to make your monthly payments during that period, but the interest will continue to accrue and be added on to the end of your mortgage.

This is only a temporary solution and is not something the building societies like doing, so don't expect them to roll over your interest payments every January and February just because you have had a good Christmas.

It really only works with a repayment mortgage. Although the building society or bank could do it with an endowment mortgage you would still have to pay the endowment premiums to the insurance company.

Long-term problems

These are a different animal altogether. If there is little hope of your ever being able to pay your mortgage properly in the foreseeable future – perhaps if you have lost your job and have little likelihood of getting another – rolling over the monthly payments for a short while won't help. It will only put off the day when you have to face the problem. Rather than let the debt continue to pile up the bank or building society will expect you to take some action.

State help may be the answer. If you qualify for Income Support then half the interest on your mortgage will be paid for the first sixteen weeks that you are unemployed. Thereafter the whole amount will be allowed. The DSS has made arrangements with banks and building societies to accept half the interest for the sixteen weeks and reschedule the arrears, which will then be taken into account by the DSS when paying all the interest on your mortgage. No capital repayments will be made by the DSS.

Income Support is a means-tested benefit and you won't qualify if your capital is over £6,000.

In the end, if you can't pay your mortgage you will have to sell your house to repay it. And the sooner you face that fact the better. The longer you leave it, the more the interest will increase the debt so that when you do sell up there will be less of the capital sum left over for you – unless you are lucky enough to find that house prices in your area rise faster than interest charges.

If you would like advice and guidance, ring the Housing Debtline on 021 359 8501.

PAYING OFF YOUR MORTGAGE

Psychologically it is a wonderful feeling to pay off your mortgage. Financially it doesn't always make much sense.

If you have an endowment mortgage let it run its full course. The endowment policy really comes into its own in the last few years and builds up dramatically.

So paying it off a few years early just to feel better could lose you thousands of pounds.

In contrast, it is often better to pay off a repayment mortgage in the last few years of its life if you can afford to.

By the time you get to the last three or four years of a repayment mortgage you will find that your monthly payments are mostly capital and very little interest. As you only get tax relief on the interest part of the payment the tax relief is down to virtually zero so you are probably paying more in interest than you could be getting as investment income on your savings.

Warning
Remember that once you have paid off your mortgage you can't get that money back again. As a mortgage is the cheapest form of borrowing there is (because of the tax relief on the interest) think long and hard about whether you can afford to use your savings to pay off the mortgage before you do it. And never take out another loan to pay off the mortgage.

15 Selling Your Home

When to sell · Should you use an estate agent? · What to do about gazumpers · Moving house versus building an extension

See also

- *Chapter 13, Buying a Home*
- *Chapter 14, Mortgages*

When it comes to selling your home you have one great advantage. You've sailed into these waters before when you bought your house or flat. And whether it all went smoothly, or you had a stormy ride, you'll have learnt a lot from the experience.

SELL FIRST OR BUY FIRST?

Most people in Scotland buy their new home before they sell the one they are living in; in England and Wales it tends to be the other way round – you find a purchaser first and then look for somewhere to buy. Unless you have a particular reason for doing so, I'd advise you to stick to the conventions of the country you are living in. Bucking the system can cost you time and money.

It means of course that in Scotland you will almost certainly have to pay for a bridging loan for the period when you 'own' two houses.

In England and Wales you will get involved in the tedious process of a 'chain'. That is the series of transactions that builds up from the first-time buyer Mr Green, who puts in an offer for Mr Brown's flat, who puts in an offer for Mrs Stone's maisonette, who puts in an offer for Mr and Mrs Smart's town house, who put in an offer for old Mrs Hogg's detached nine-bedroom Edwardian villa who is moving to live with her daughter. And they all intend exchanging contracts on the same day. It can take months for the chain to build up, it inevitably breaks down at least once and you'll probably be sick with worry by the time you get anywhere near your new property. But that is the system and it more or less works.

If, in England, you buy your new home before selling your old one, and then sell your house into a 'chain' that falls down, you could have your expensive bridging loan like an albatross round your neck for months if not years.

A growing trend nowadays, in the south, is to sell first, move into rented accommodation and then buy as a cash buyer. Not an easy course, though, if you have children at school.

When to sell

There is no best time to sell your house, when you are bound to get a higher price. The right buyer could come along at any time. However, fewer people think about moving house at Christmas. The busiest time for moves is the school summer holidays, so spring is the peak time for looking at and buying new homes. Try to put your house on the market in March, or, failing that, September.

But if you are using a photo to illustrate the property, take the picture at the most picturesque time of year, probably the early summer when the sun is shining and the garden, if you have one, is looking its best.

Estate agent or sell it yourself

Most people in England and Wales and an increasing number of sellers in Scotland use an estate agency. For a percentage of the value of your house – usually between 2 and 3 per cent – the estate agent will handle the sale for you. He'll value your property, draw up details to give to interested buyers, deal with phone calls and even take clients round your home. He'll do the advertising, put a board outside and negotiate the price with the potential new owner.

If you decide to go it alone you'll save yourself the estate agent's commission, which could be up to £1,300 on a £60,000 house. Because you are only selling one house you'll put more effort into selling it than the agent might, as he has a lot more properties on his books.

Against that you will have to do your own haggling with any buyer and pay for your own advertising. You won't have the use of the estate agent's client base, nor his shop window to attract buyers. And any buyer, knowing that you are not using an estate agent, might expect you to cut the price of your property by half the potential commission.

In the long run you might be better to use an estate agent *and* try to sell the property yourself. That way you get the best of all worlds. If he sells it for you, pay up. If you sell it yourself, don't pay up. But watch out when you sign the estate agent's contract. Don't muddle 'sole agency' and 'sole selling rights'.

Sole agency means that is the only estate agent you are using. *Sole selling rights* means that the estate agent gets the agreed commission even if you sell it yourself, privately. Avoid that like the plague.

Tip
Check what you are paying for if you use an agent offering cheap commission rates. You might find you have to pay an additional sum for putting a 'For Sale' sign outside your property and for all the advertising the agent does for you. That will be particularly galling if your house does not sell and you still have to foot the agent's bill.

One estate agent or several

If you decide to go down the estate-agent route, your next decision is to decide which agent to use, and whether to use one, or several.

Look around your area at the 'For Sale' boards and see which agent is the most popular, and which boards seem to change from 'For Sale' to 'Sold' most frequently. That is the agent to go for.

It is up to you whether you just use one agent or several. If you use just the one, the commission rate will be about ½ per cent lower. If you use several you'll have to pay more to the one who sells it, but you get potentially more people to look round your property.

But don't have more than one 'For Sale' board outside your house, as it looks as if you are desperate to sell. That would encourage any potential buyer to knock your price down. It is also against local planning rules.

Property shop

You can use a property shop instead of, or as well as, an estate agent. For a set fee of around £100, the details of your house will be put on to a computer and matched with people looking for properties. The details will stay on the computer list until the house is sold but the fee has to be paid whether or not the house is sold through the property shop. If you can afford it it is probably worth doing even if you are using an estate agent as well. Your local paper will have details of the property shops in your area.

Gazumping

Whether you like it or not, gazumping happens in England and Wales when house prices are rising. If you have an offer on your house, but you haven't yet exchanged, and someone comes along and offers you more, what should you do?

Morally you should refuse to accept the new higher offer. But you are in this to get the highest price for your house, so don't feel too badly about accepting. You could always salve your conscience by refunding the original buyers any money they have spent on surveys, and asking if they would like to match the new offer.

But watch out. Make sure the difference in price is worth it, or you could lose a firm sale for the sake of a slightly higher one that fades away.

Fixtures and fittings

You can take with you when you move as many or as few of the fixtures and fittings as you like. But make sure the buyer knows what is going. You can include the carpets and curtains in the price as a selling point, or negotiate a separate deal for them with the buyer or leave them for nothing. What you can't do is say that you are going to leave them when you move, and then take them with you.

And remember the garden. If you have favourite plants that you want to take with you, tell the buyer in advance. You can't sell a house with a leafy mature garden and then uproot the lot and move it with the furniture.

Removal day

Whether you use a removal firm which packs and shifts all of your goods and chattels or hire a van and do it yourself, make sure you do something about insurance.

It is not the end of the world if you drop and break a mirror or damage one of your kitchen chairs but if the van containing all of your furniture crashes badly, or is stolen, you could lose thousands of pounds worth of goods.

If you use a reputable removal firm they will either have proper insurance or advise you to get it. But it is always worthwhile checking with the company which does your house contents insurance. If you are moving and using a reputable removal company they will often cover the move at no extra cost or for just a few pounds.

Remember to tell your insurance company that you are moving house or your house contents cover could become invalid.

Sell or extend?

If you are moving house because you want a larger property, you might be put off by the costs of selling and buying or your inability to sell. Extending your current home may seem like a less costly and easier alternative.

You might also think that you will increase the value and marketability of your house that way too. Not always.

If you add another room to your house by extending over, say, the garage, then you may add real value to your property. But if you build your extension on the side or back of your house, and lose part of the garden in the process, you may not gain as much value. If it suits you to do that, well and good, but remember it might not suit future buyers.

Home improvements

Home owners spend around £10 billion in home improvements every year. New kitchens are put in, wardrobes fitted, lofts converted, patio doors added and double glazing installed. But how much of that money has increased the value of your home?

If you do something to your home to suit yourself, enjoy it. But don't expect the next owners to like it, or indeed to pay for it.

The only improvement that you can make to your home that will definitely increase its value by more than the money you have spent is installing central heating. Everything else you do at your own expense. It may be a good investment but it is more likely that you will not get your money back if you are doing the improvement solely to increase the value or marketability of the property.

Indeed something like building a swimming-pool in the back garden can actually decrease the value of the property because the cost of the upkeep of a pool may put buyers off, particularly if they would rather have had the garden. And a swimming-pool is almost as expensive to get rid of as to install.

And whatever you do, don't over-improve your house. A five-bedroom mansion in a street full of two-bedroom terraced houses won't sell for anything like the price you'll expect to get.

16 Living Together and Getting Married

Your financial rights if you are not married – to wed or not to wed · How marriage affects your tax bill

See also

- *Chapter 7, Understanding Income Tax*

LIVING TOGETHER

Not so many years ago, a woman living with a man who was not her husband was regarded as 'scarlet'. She might be cruelly described as a mistress (if not a whore), living over the brush, or living in sin.

Times change. It is now definitely acceptable to live together, unwed. And the terminology has taken a turn for the better: cohabiting, entering into a premarital relationship, even having a toy boy!

But whatever you want to call it, living together can cause a financial tangle.

Let your heart rule your love life, but make sure your head has a say when it comes to sorting out the money.

No matter how long you live together, or how much your relationship is like a marriage, in the eyes of the law and the tax man, a couple are not married unless they go through a civil or religious ceremony, and they will not be treated as such. So a few hours spent sorting out the finances can pay huge dividends if anything goes wrong. This is particularly important if there are any children.

Common-law wife

In England and Wales there is no such person as a common-law wife – legally the term does not exist. So you will be treated by the law as a single person. However, should your relationship break up after many years, during which the woman brought up children and contributed to the costs of running the home, then that will be taken into consideration by the courts (if it comes to that). But the woman will only get a share of what she helped to pay for or increased the value of, say by helping to modernize the house; she won't be able to claim a share of any assets that are in the man's name. And vice versa. 'Palimony' is not accepted by the courts either.

In Scotland there is such a thing as 'marriage by cohabitation, and habit and repute'. That means, in layman's terms, that a common-law wife does exist legally.

You have to establish that you are living together and that everyone accepts you as husband and wife. For example, a woman might be expected to change her name by deed poll to the man's name.

On the death of the man, she would have to go to the Court of Session and apply for a Declarator of Marriage. That would then entitle her, legally, to all the financial benefits a widow would have got.

You cannot get a Declarator of Marriage, or be accepted as a common-law wife, if a real wife exists, or if the woman is not divorced from a previous husband.

If you intend trying to claim anything as a common-law wife – get professional advice. Either see a solicitor or go to a Citizens' Advice Bureau.

Income tax

To the Inland Revenue a person is either married or not married. They can be divorced, widowed, single, legally separated or even dead. But one thing they cannot be is living together. That is not a concept that the tax man accepts. So if you are living together, even if you now share the same name and have children, you will be treated as two single people and taxed as such.

Since 1990, everyone has been taxed independently, whether they are married or not. The tax rules have been rewritten so that no woman loses out, financially, by getting married. A working couple, living together, will both get a single person's tax-free allowance. If they marry, they will also get a married couple's allowance. So that's a disadvantage to the unmarried.

If you have children you can claim an additional tax-free allowance as a single parent (see p. 70) that you won't get if you are married.

However, if there are two or more children, a divorced or separated couple can do better. If one child lives with the mother and the other with the father, the parents can each claim an additional tax-free allowance. Unmarried couples living together get only one allowance between them.

For full details see p. 67 in Chapter 7, Understanding Income Tax.

The Department of Social Security takes a different view. In their eyes you are either living alone or not living alone. So if you share your life with someone, however temporarily, the DSS will take your new circumstances into account and change your benefits accordingly.

Maternity

Being married makes no difference. Any claims you make for maternity benefit or statutory maternity pay are judged on the amount of National Insurance contributions you yourself have made. Neither your husband's nor your partner's contributions are taken into consideration.

Pensions

A married woman will do better than her single sister. Both can earn a state pension in their own right, but if the women have not earned that right by paying enough National Insurance contributions of their own, the married woman might still qualify through her husband's contributions. The single woman will have no rights to her cohabitee's contributions.

When it comes to the company pension the rules for unmarried women are quite complicated.

Unmarried women and company pensions

If the man dies *before* retirement, only a widow will get the widow's pension in most schemes. No marriage certificate, no pension, is the ruling in many company schemes. Some are more flexible and will pay out to a financial dependant, but if there is also a genuine widow at least part of the pension will go to her.

A lump-sum payment would be at the discretion of the pension fund trustees and an unmarried partner might get it if there is no more deserving claim, such as an old disabled mother, or sister, but some pension schemes pay the lump sum into the estate of the dead employee and it would then be distributed according to his will.

The third benefit: a refund of the man's own contributions would be paid into his estate and disposed of according to the terms of his will or, occasionally, at the discretion of the trustees.

If the man dies *after* retirement, again the widow's pension will only go to a widow. Many schemes nowadays also have a five-year guarantee on the pension so that if the man dies in the first five years of his retirement then the full pension is paid for that period though this is nearly always paid as a lump sum. That payment will either be at the discretion of the trustees, in which case it could be paid to an unmarried partner, or it might be paid into the member's estate.

But even if the man has been separated from his wife for many years and has a long-term relationship with his partner, on his death it is the wife, not the partner, who will get the widow's benefit. That is the law and the trustees have no discretion over most of the payments.

Tip

If you want your partner to benefit, where she can, from your pension scheme, make sure you nominate her on an 'Expression of Wish' form. These are held by the trustees and used when deciding where the lump sum should be paid. For tax reasons, the trustees are not obliged to follow your wishes, but they generally do.

Unmarried men and company pensions

Unmarried men now have the same pension rights – or lack of them – as unmarried women. A European Court judgement found it illegal to discriminate between the sexes in pensions which means that men and women now all have the same rights on pensions, benefits from them and retirement age.

Pensions and dependants

If you are living with someone to whom you are not married, there is something you can do, on retirement, to secure your pension for them. You can opt to take a lower pension, so that a percentage continues to be paid to your dependant after you die and until your dependant's death. It doesn't have to be your cohabitee. If you live and financially support a sister or parent you could name them as your financial dependant. The figures are very sensitive to the age difference. If your dependant is much older than you, then they would get more than the table below shows. If they are much younger, they would receive much less.

For example, a man of sixty-five with a wife or dependant of sixty-two has a pension of £100 a week. When he dies, his pension will die with him. If he wants it to continue on and be paid to his wife (or dependant) he can do this, but at a cost, as the example shows.

Proportion of pension received by wife on death	*Husband's weekly income now*	*Wife's weekly income on death*
Full pension	£78	£78
$\frac{2}{3}$	£84	£56
$\frac{1}{2}$	£88	£44
$\frac{1}{3}$	£91	£30

Death

Morbidly, it is on the death of the man that the unmarried woman really does miss out unless preparations have been made. A couple living together must write wills, otherwise the partner who survives could lose all rights to the other person's assets. If you haven't got a will already, do something about it now. Start by reading Chapter 29, Making a Will.

You may want all you own to go to your partner but unless you say so in a will, it may fall to that third cousin in Scunthorpe whom you never liked. The surviving partner would have to apply through the courts for a share in the inheritance and that is just the sort of hassle you don't want after the death of someone you love.

And if you are rich enough to be leaving more than £150,000 (tax year 1992–3) to your surviving partner, you will also be leaving a tax bill. Only husbands and wives can transfer more than £150,000 to each other on death without incurring a bill for inheritance tax.

The unmarried woman doesn't qualify for widow's benefit either.

Divorce

If you don't marry, you can't divorce, so sorting out the finances after a bitter break becomes a lot more difficult for the unmarried couple.

Neither partner has to pay maintenance to the other partner. If there are children involved and the man doesn't voluntarily help with their upkeep, the mother has to apply to the courts for what's known as an affiliation order; this is a nuisance and it can be embarrassing, so few mothers do it.

GETTING MARRIED

Almost 400,000 people tripped down the aisle or tied the knot in a registry office in 1989 – and very few of them were influenced by tax considerations when taking the decision to get married.

None the less, the main difference between being married and unmarried, in money terms, is in your tax affairs.

Nowadays there is no 'best time' of the year to get married; whenever you marry, the Inland Revenue treats the couple as married as soon as they tie the knot.

The married man

In the year that he marries, the man will become a tax husband immediately. He will get the part of the married couple's tax-free allowance that he is due. For example, if he marries halfway through the year, he'll get half of the allowance; if he marries one month into the tax year he'll get eleven months' worth of the married couple's allowance.

If the marriage takes place on 6 April, the first day of the new tax year, the couple will be treated as being married for the whole of the tax year.

To claim the married couple's allowance, write to the Inland Revenue, using Form 11PA, telling them what day you got married. Your tax inspector will then change your tax code to take into account the increase in your tax-free allowance.

The married woman

The personal allowance a woman receives on her income tax will be the same whether she is married or unmarried. However, after she ties the knot she is entitled to a share of the married couple's allowance.

It is given automatically to the husband, but if the woman wants half of it, to offset against her tax, she only has to ask the Inland Revenue. And she doesn't need her husband's permission to claim it. If the woman is a higher-tax payer than her husband (or is a taxpayer while he is not) it is in both of their interests for her to claim it all. In which case they would both have to agree to the transfer.

Mortgage relief

If both partners have homes and both are getting tax relief of up to £30,000 each on the mortgage interest, then they will continue to get it for up to twenty-three months after they are married.

Assuming the woman is moving in with the man, the Inland Revenue will allow her tax relief on the mortgage for the tax year in which she is married and the following one, unless she sells up. And if she is really having difficulty selling her former home, her tax man might even extend the period for mortgage relief until she does sell. The tax situation would be reversed if the man moved in with the woman.

Thereafter, the couple will only have one £30,000 allowance between them.

17 Having a Baby

What you can claim when you are pregnant · Taking time off work · Your right to your job back

See also

- *Chapter 19, Children and Money*

THE COST OF HAVING A BABY

Around three quarters of a million babies are born in this country every year – over half of them to first-time mums. They will bring joy and happiness, but they will also bring additional costs to the household budget.

I have really only had two major financial disasters in my life so far – one is called Laura and the other Jamie.

If the mum-to-be has been working up till her pregnancy, and now plans to give up, the financial strains will be increased enormously. Losing part of your family income, just when the spending increases, always causes money problems.

Insuring against twins

If you're worried that you might be having twins, or more, you can insure against the extra costs. Providing you're not on any fertility drugs and that you fill in the proposal form within nine weeks of conception – and before any scan – you should get insurance. The premiums rise as you get older or if there is a history of twins in the family.

For a twenty-five-year-old with no history of twins in the family it will cost around £2 per £100 of insurance (that means if you want a pay out of £2,000 if you have twins, the premium would be £40), while for a thirty-eight-year-old whose mother was a twin the cost rises to £10 per £100 insured. There is usually a minimum charge of around £50.

Several of the large insurance companies, such as General Accident, offer this sort of insurance and it is worth phoning several to compare quotes.

STATE HELP BEFORE THE BABY

All mums-to-be are entitled to free prescriptions and free dental treatment for the nine months of their pregnancy and for a year after the baby's birth.

Fill in Form FB8, available from your doctor or midwife, and send it off to the address given on the form as soon as you know you are pregnant in order to claim your free prescriptions.

And tell your dentist when you next go for dental treatment so that you don't have to pay. But do ensure first that you are being treated on the NHS, not privately, otherwise you'll continue to get the bills.

Help for mums-to-be on low incomes

Since the £50 Maternity Grant was phased out, there is no longer a statutory lump sum given to all expectant women. Pregnant women on low incomes can apply for help from the DSS Social Fund. It will pay up if the woman or her partner is on Income Support or Family Credit, but not otherwise, and is a one-off sum through the Social Fund maternity payment. What you get depends on the savings the family has. The most you'll get is £100, but if you have savings of more than £500, that will be reduced pound for pound. So if your savings reach £600 you'll get nothing at all from the Social Fund.

Pregnant women on Family Credit and Income Support also qualify for:

- free milk – up to seven pints a week
- free vitamins
- help with fares to and from hospital for antenatal visits
- a lump sum to help with buying things for the new baby such as the cot, pram, nappies and a bath, and buying maternity clothes and shoes – these are discretionary
- help, or additional help, with rent and council tax.

To claim, you have to fill in the appropriate DSS form, and these are listed at the end of this chapter.

Help for working mums-to-be

Statutory Maternity Pay

Pregnant women who are working can claim Statutory Maternity Pay. The system is operated by the employer but the woman doesn't have to intend returning to work to get the money.

To qualify the woman has to:

- have been employed by the same employer for at least the six months running into the twenty-sixth week of pregnancy, without a break
- have been earning over £54 a week (the lower earning limit for National Insurance contributions) during the last two months up to the twenty-sixth week of pregnancy.

Statutory Maternity Pay lasts for eighteen weeks and will start to be paid after you have stopped work. For the first six weeks of the eighteen you get 90 per cent of your basic weekly salary provided you have worked for at least two years (five if you work part time); for the other twelve weeks it is a flat rate, currently £46.30 a week, no matter how much you previously earned or paid in National Insurance contributions. If you have worked for the same employer for more than twenty-six weeks but less than two (five) years you can get the flat rate throughout the SMP period.

You do have some flexibility as to which eighteen weeks you claim your Statutory Maternity Pay for. There is a core period of thirteen weeks that you have to take from six weeks before the baby is due until six weeks after the expected week of delivery. But you can suit yourself when you have the other five weeks. You can take it all before the birth, all after, or split it around the core period.

You do have to give your employer three weeks' notice of your date of leaving, and fill in the appropriate Department of Social Security form, listed at the end of the chapter. You also have to give him your maternity certificate (Form MAT B1).

Maternity Allowance

If you are self-employed, have recently changed jobs, or for some other reason don't qualify for SMP, you may be able to claim the other main state maternity benefit – the Maternity Allowance.

To qualify you must have been employed or self-employed and have paid the standard rate National Insurance contributions for at least half of the year ending in the fifteenth week before your baby is due. You should receive £42.25 a week for eighteen weeks.

Again you have some flexibility as to which eighteen weeks you claim for, and the rules are the same as those which govern Statutory Maternity Pay. You cannot receive payment for any period you are working.

You claim your Maternity Allowance from your local social security office. Even if you haven't worked recently, or are still at school or college, you may be eligible for at least part of the allowance, so it is worth making a claim.

TIME OFF WORK

If you want to return to work after the baby's born, you'll be entitled to your job back if you've worked full time for the same employer for at least two years, or part time for five years, and continued to work for the first twenty-nine weeks of your pregnancy.

But the conditions are complicated. You must write to your employer at least three weeks before stopping work telling him you're expecting a baby, when it's due, and that you'd like to return to work afterwards. This letter doesn't bind you to returning to work if you change your mind later, but it binds the employer to having you back.

Seven weeks or so after your baby is born, your employer will write asking if you still want your job back, and if you do you must reply within three weeks giving him a date when you'd like to restart. If you don't receive a letter from your employer, write to him by the ninth week after the birth, confirming your intention to return. And you must be back at work within twenty-nine weeks of the birth. If you don't fulfil all of those conditions, you lose your right to your job.

If your employer refuses to hold your job open for you, you can claim unfair dismissal.

You'll be entitled to any pay increases that have been made while you've been away. Those missing months won't count towards your pension but other benefits such as redundancy shouldn't be affected.

You're also entitled to paid time off for antenatal appointments.

Some companies, particularly large ones with organized personnel departments and, perhaps more importantly, powerful unions, are more generous than the legal minimum. They often allow longer maternity leave and will continue to pay you for longer than the government insists. But you still need to comply with their rules if you want to get your job back.

STATE HELP AFTER

Child Benefit

Once the baby is born you're entitled to Child Benefit. This is a tax-free cash sum, currently £9.65 a week for the first child, and £7.80 a week for each of the others, paid to the mother until her children reach the age of sixteen, or up to nineteen if they remain in full-time education.

> **Tip**
> Apply for your Child Benefit claim form before the baby is born and fill in as much as you can of it. You can't complete it and send it off until you have a birth certificate. But you will have very little time when you arrive home with a new-born baby and the longer you put off sending in the form the longer it will be before you get your Child Benefit book.

One-parent Benefit

You can apply for One-parent Benefit if you're bringing up a child on your own. You get this on top of Child Benefit and it is a tax-free weekly sum of £5.85, but unlike Child Benefit it is only paid once, no matter how many children you have.

> **Tip**
> Don't delay claiming any of your money from the DSS – whether it is Child Benefit or One-parent Benefit. The later you send in your forms, the later the money will be in coming. And if you delay too long, without good reason, you could lose out.

CLAIMING THE BENEFITS

To make a claim or to find out more about what you're entitled to, the following DSS forms and leaflets will be of help:

FB8, 'Babies and Benefits'
MA1, 'Maternity Allowance'

Checklist: pregnancy and birth		
When	*What to do*	*Why*
Once you are pregnant	• Ask GP for Form FB8	Free prescriptions
	• Tell dentist	Free dental work
	• Tell DSS if on FC	Check right to free milk and vitamins and hospital fares
	• Tell employer	Paid time off for antenatal visits
17 weeks before birth	• Fill in Form MA1 • Ask GP or midwife for Form MAT B1	To support your claim for Maternity Allowance or to give to your employer if you can get SMP
11 weeks before birth	• If on Income Support or Family Credit ask DSS or antenatal clinic for Form SF100	To claim a maternity payment from the Social Fund
Before you stop work	• Tell employer three weeks before you stop when baby is due and that you intend returning	To protect right to SMP and to return to work
As soon as possible after birth	• Register baby's birth	To get birth certificate
	• Apply for Child Benefit	To get Child Benefit
	• Check low-income benefits	To see if extra child qualifies you for extra benefits
6 weeks after birth (3 in Scotland)	• Register baby	Latest date
9 weeks after birth	• Write to employer saying you intend returning to work	To protect right to return
3 weeks before returning	• Write to employer giving date	To protect right to return
29 weeks after birth	• Return to work	Or lose right to return

SMP1, 'Statutory Maternity Pay'
N117A, 'Maternity Benefits'
FC1, 'Family Credit'
H11, 'NHS Hospital Travel Costs'
P11, 'NHS Prescriptions: How to Get Them Free'
MV11, 'Free Milk and Vitamins'
RR1, 'Housing Benefits – Help with Rent and Poll Tax'

CH11, 'One-Parent Benefit'
CH1, 'Child Benefit'
SB16, 'A Guide to the Social Fund'

18 Insurance

Buildings insurance · House contents insurance · Motor insurance, including car, motor bike and caravan · Holiday insurance · Permanent health insurance · Private medical insurance · Legal expenses insurance · What they are, what they cost, what they cover and what they exclude

See also

- *Chapter 6, Life Insurance*

You could probably spend just about everything you earn paying for insurance. Insuring yourself, your house, your car, your job, your holiday, your health, your pets – even your contact lenses. Or you could take out no insurance at all and just hope that nothing bad ever happens to you.

For most of us, the problem is finding the right balance between those two extremes so that we don't spend all our extra money on insurance but we do take out enough cover.

If you never have to make a claim on your insurance policies you will grudge every penny you spend on them – rather than thank your lucky stars that you have had no accidents. If you do claim, and you haven't been careful when you took out the policy, then you will probably find that the one thing you are claiming for is not covered. So it is worthwhile spending a little time making sure you have the right insurance – and at the right price.

There are two types of insurance you must have:

- house building insurance, if you have a mortgage, because the lender will insist on it
- motor insurance, if you have wheels, because that is the law.

There are several more which you should have:

- house contents insurance
- travel insurance if you go abroad
- mortgage protection.

And there are plenty of others which are optional and which you should take out if you think you'd like that type of cover, such as legal expenses insurance, permanent health insurance, medical insurance and redundancy insurance.

There is a fourth type of miscellaneous insurance that you can buy if you want cover against more unusual things such as vets' fees, having twins or rain spoiling your village fête.

Life insurance, being a form of assurance, has a chapter to itself. See Chapter 6.

HOUSE INSURANCE

There are two types of house insurance:

- buildings insurance
- contents insurance.

It is quite easy to work out which one insures what. If you think of everything you would take with you when you move home, then that is what's covered by contents insurance; everything you leave behind is covered by the buildings insurance. So the bricks and mortar, fitted wardrobes and wallpaper would come under buildings; furniture and all your personal belongings, under contents.

Tip
Have the building and the contents insured by the same insurance company so that if you have a serious accident such as a fire or a burst pipe and both the fabric of the house and some of the contents, for example the wallpaper and the furniture of one room, are damaged, then at least you only have to deal with one insurance company and one loss adjuster. There will be no haggling over who pays for the cupboard that may or may not be fitted.

Buildings insurance

Most buildings insurance policies cover loss or damage to your house or flat by any one of the following disasters: fire, theft, escape of water from tanks or pipes, oil leaking from fixed heating systems, storm, flood, riot or malicious acts, explosion, lightning, impact by aircraft, vehicles or animals, earthquake, subsidence, landslip or heave, falling trees or falling aerials.

Most buildings policies also include cover for alternative accommodation so that if your home is so badly damaged by a fire or a flood that you cannot live there while it is being renovated the insurance company will pick up the bill. Not all of the bill – there is a maximum of 10 per cent of the sum insured.

You can insure against the glass in your home being broken – the windows, patio doors, skylights – and baths, basins and toilets. You can also get cover for accidental damage to underground gas and electricity cables and oil and water pipes. But the more cover you have, the more you will have to pay.

You won't be covered under any policy for general wear and tear, woodworm, dry rot or gradual deterioration of your home.

Accidental damage can be added to most buildings policies, though you will have to pay extra for it. It means that the damage caused by you putting your foot through the ceiling while working in the loft is covered.

If you have a mortgage, the bank or building society which gave you the loan will insist that you have buildings insurance. If you don't have a mortgage you should buy this type of insurance anyway.

Tip

You don't have to go for the buildings insurance that your bank or building society particularly recommends. Provided you insure through a reputable company, the policy will be acceptable to the mortgage lender, though you may have to pay an administration fee. So if you have a good deal with the insurance company which covers your house contents, move your buildings policy there as well. Tell your bank or building society before you change.

What is covered

Apart from your home, most buildings policies will also cover any permanent structures in your garden such as a garage, swimming pool, shed, drive, walls and sometimes fences. Walls, gates, and fences are not always included in the cover if the damage is caused by a storm.

Cost

Buildings insurance costs much the same from most insurance companies, so it is not one to shop around for. What you pay depends

These costings do not apply to all types of property – see notes overleaf
JANUARY 1992 costings – £/ft² gross external floor area

		PRE-1920			1920–45			1946–79			1980–DATE		
		LARGE	MEDIUM	SMALL	LARGE	MEDIUM	SMALL	LARGE	MEDIUM	SMALL	LARGE	MEDIUM	SMALL
DETACHED HOUSE	Region 1	67.50	72.50	73.00	64.50	68.00	69.50	53.50	58.00	60.00	53.00	52.50	56.50
	2	61.00	65.50	65.50	58.00	61.00	62.50	48.50	52.50	54.00	47.50	47.00	51.00
	3	58.00	62.00	62.50	55.00	58.00	59.50	46.00	50.00	51.00	45.50	45.00	48.50
	4	55.00	59.00	59.00	52.00	55.00	56.50	43.50	47.00	48.50	43.00	42.50	46.00
	Typical area ft²	3450	1700	1300	2550	1350	1050	2550	1350	1050	2400	1400	950
SEMI DETACHED HOUSE	Region 1	65.50	67.00	67.50	70.00	67.50	67.50	50.50	53.50	57.00	58.00	55.50	60.00
	2	59.00	60.50	60.50	63.00	60.50	61.00	45.50	48.00	51.50	52.50	50.00	54.00
	3	56.00	57.50	57.50	60.00	57.50	58.00	43.50	45.50	48.50	49.50	47.50	51.00
	4	53.00	54.50	54.50	56.50	54.50	55.00	41.00	43.50	46.00	47.00	45.00	48.50
	Typical area ft²	2300	1650	1200	1350	1150	900	1650	1350	1050	1600	900	650
DETACHED BUNGALOW	Region 1	The chart does not cover pre-1920 bungalows, as few such properties were built			67.50	62.50	65.00	57.00	58.00	60.50	59.00	59.50	61.00
	2				61.00	56.50	58.50	51.50	52.00	54.50	53.00	53.50	55.00
	3				58.00	53.50	55.50	48.50	49.50	52.00	50.50	50.50	52.50
	4				54.50	51.00	52.50	46.00	47.00	49.00	47.50	48.00	49.50
	Typical area ft²				1650	1400	1000	2500	1350	1000	1900	950	750
SEMI DETACHED BUNGALOW	Region 1				68.50	66.50	64.00	54.50	55.50	60.00	57.00	64.00	67.00
	2				62.00	60.00	58.00	49.00	50.00	54.00	51.00	57.50	60.50
	3				58.50	57.00	55.00	46.50	47.50	51.50	48.50	55.00	57.50
	4				55.50	54.00	52.00	44.00	45.00	48.50	46.00	52.00	54.50
	Typical area ft²				1350	1200	800	1350	1200	800	950	550	500
TERRACED HOUSE	Region 1	71.50	70.00	70.50	70.00	69.50	69.50	50.50	54.50	60.50	56.50	58.50	58.00
	2	64.50	63.50	63.50	63.00	63.00	62.50	45.50	49.00	54.50	51.00	53.00	52.50
	3	61.00	60.00	60.00	60.00	59.50	59.50	43.00	46.50	51.50	48.50	50.50	50.00
	4	58.00	57.00	57.00	56.50	56.50	56.00	41.00	44.00	49.00	46.00	47.50	47.00
	Typical area ft²	1650	1350	1050	1350	1050	850	1650	1300	900	990	750	650

Regions

1 London Boroughs and Channel Islands*

2 South-East
Bedfordshire, Berkshire, Buckinghamshire, Essex, Hampshire, Hertfordshire, Kent, Oxfordshire, Surrey, East Sussex, West Sussex

3 East Anglia
Cambridgeshire, Norfolk, Suffolk
North-West and Scotland

4 East Midlands, Northern, South-West, West Midlands, Yorkshire and Humberside, Northern Ireland† and Wales
All other counties

*Building costs in the Channel Islands are affected by local conditions and may vary from prices in this band. You should seek local advice.

†Building costs in Northern Ireland are considerably lower than in the rest of the UK and may be 10-30% below the costs given for Region 4.

How much would it cost to rebuild your home?

on the sum insured, and that in turn depends on where you live and the size of your home. You don't insure for the market value of the property, but for the rebuilding costs. So in the south-east of England, where house prices are high, the rebuilding cost will probably be less than the market value. In the north-east, where house prices are low, the rebuilding cost could be higher than the market value.

The price for a standard brick-built house is around £1.80 to £5.00 per £1,000 insured. With almost biblical overtones, you pay more if your house is built on clay, to take account of the risk of subsidence, and less if it stands on solid rock. To work out how much to insure, first measure your house; then, using the chart, identify your type of home and multiply the cost per square foot by the number of square feet in your house. Then add on an allowance for the garage, shed and fences. When you get a figure of so many thousands of pounds, multiply that by £3.60 per thousand and that gives you the premium you should expect to pay. So a large Victorian semi in the south-east would cost around £530 a year to insure, a medium modern detached bungalow in Yorkshire, around £180 a year.

Excess

To try to keep small claims to a minimum, most buildings policies operate what they call an 'excess'. That means you pay the first part of any claim yourself.

So, if you are claiming for damage caused by a burst pipe, for example, you will probably have to pay the first £15 or £20 – your policy will show the exact amount.

The exception is subsidence, where the excess is £1,000.

> **Tip**
> Don't try to cut the cost of your premium by underestimating the size of your home. Although very few people have their house totally destroyed by fire or flood, any claim that you make will be scaled down by the insurance company if you have underestimated, and they may not pay at all if they think that you have cheated them on the premium by deliberately under-insuring.

Making a claim

Tell your insurance company as soon as possible after the damage is done – and don't start repairs until you get the permission from the company. However, you can go ahead without permission and do temporary work such as putting tarpaulin over a leaking roof after a severe winter storm to prevent further damage – that cost would also be covered by the insurance company.

Once you obtain the go-ahead from your insurance company, get the work done and send in the receipts. Don't use anyone that can't supply you with a receipt or the insurance company won't pay up, because you cannot prove how much it cost.

> **Tip**
> Take photographs of the damage and prepare a list of the damage to your property. It will prove very useful when arguing with the loss adjuster.

House contents

Insuring the contents of your home is a much more complicated affair. For a start, no one makes you do it. It would be a brave person who could afford to insure their belongings yet chooses not to do it, but one in four households don't bother, some of them through apathy, others simply because it is becoming so expensive.

However, once you understand exactly how contents insurance works, you could choose one of several types of policies which will give you a cheaper type of insurance. Admittedly it will also cut the level of cover but I think that you are better to have something than nothing.

What is covered

Basic. All house contents insurance policies cover you for the same disasters as buildings insurance – that is: fire, theft, escape of water from tanks or pipes, oil leaking from fixed heating systems, storm, flood, riot or malicious acts, explosion, lightning, impact by aircraft, vehicles or animals, earthquake, subsidence, landslip or heave, falling trees or aerials.

Accidental damage. If you have expensive pieces of furniture in your home and small accident-prone children, you can extend your insurance cover by taking out an accidental damage policy to cover breakages and damage that you or anyone else does in your home. Even this won't cover you for every cup you drop or glass that gets broken. Some accidental damage policies will not cover damage caused by your pets either.

All risks. If you travel a lot you can take out an 'all risks' extension which will cover valuable items lost, stolen or damaged outside your home. Things like fur coats, jewellery, watches, spectacles and cameras which you use away from your home will then be covered wherever they are lost, stolen or damaged.

What you pay for accidental damage and all risks insurance will depend on the value of the items you are insuring. Ask your insurance company for a quote.

The cost of basic house contents insurance

There are two levels of house contents insurance:

- indemnity
- new for old.

Indemnity. This type of insurance gives you the second-hand value of your goods. The insurance company will deduct an amount for wear and tear and depreciation before paying out to repair or replace whatever has been lost, stolen or destroyed. Few people opt for this type of cover now and some insurance companies don't even offer it.

New for old. This type of insurance will replace with a new item whatever was stolen or damaged regardless of how old it was at the time.

New for old tends to be 50p to £1.50 more per £1,000 of cover than indemnity insurance.

When working out how much cover you need, you do actually have to add up the value of everything in your home. Any good insurance company will send you a ready reckoner to help you. Again, don't underestimate the value or you will get all your claims scaled down.

Tip
Some insurance companies can give you a more standard cover dependent solely on where you live and the size of your house. They call it 'bedroom' insurance because what you pay depends on the number of bedrooms you have in your house and your postcode.

Premium per £1,000 of cover

	New for old	*Accidental damage*	*All risks on valuables*
Major city centre	£18	£20	£20–40
City suburbs and large towns	£6.15	£9	£17.50
Rural areas	£3.50	£5	£12–15

Expensive items

Most insurance companies ask you to name anything that is expensive in your house, such as pictures, antiques, jewellery, hi-fi, video and computers, so that they can be insured separately. (If you rent your TV and video it is still up to you to insure them at the market value.)

Theft

If something you own that is covered by your insurance policy is stolen while it is on loan to someone else, or in your office, it should still be covered by your household policy. Quaintly, it is not covered unless it is in another *building*, and the thief breaks in, so if it is stolen from your car, or you leave it on the plane, then you won't get your money back unless that item is covered by an all risks policy.

Extras

Your household contents insurance will also cover the garden tools and lawnmowers that you keep in your shed (providing it locks). So remember to add them in when you are totting up the contents. It will usually also give you up to £1 million of personal liability cover. So if you cause an accident while out on your bicycle, or you poke someone's eye out with your umbrella and a court says you must pay them compensation, then your insurance company will pay this for you, up to £1 million.

Exclusions

All insurance policies have their own exclusion clauses but the most common relates to theft. Unless there are signs of forcible entry, your insurance company may not accept your theft claim. That means that if you never lock your front door and someone walks in and takes your TV or the video is stolen by someone you allowed into the flat, then you have no claim. However, if you are conned by a bogus water man, or just once left the front door unlocked and were unlucky enough to be burgled, you will probably have a just claim and the insurance company will pay up.

Cutting the cost of house insurance

1 Fit a burglar alarm.
2 Join a Neighbourhood Watch scheme.
3 Fit good locks to doors and windows.
4 Don't make small claims and you may qualify for a no-claims bonus.
5 Ask your union, association or any club you belong to if they do special deals for members.
6 Take expensive items off your household policy, though you must tell the company you have got them. They won't be covered so if they are stolen you lose them completely, but at least the premium will be much lower.
7 Be over fifty-five.
8 Go for indemnity rather than new for old.
9 Don't take out all risks or accidental damage cover.
10 Spread the annual premium over monthly payments, if it doesn't cost extra.

No one will be able to take advantage of all these cost-cutters, but if you choose the ones which suit you, you should lop a good few pounds off your annual premium. You might, of course, also reduce your cover, but if you're stretched financially, some cover is better than no cover at all.

Add-ons

Many insurance companies will offer 'add-on' cover on top of a basic contents policy. For example you could add on insurance for bicycles, sports equipment, the food in your freezer, the cost of replacing the locks to the house if your keys are stolen, insuring your caravan, a small boat or the vet's fees for your pets.

You may not want this insurance but if you do it is almost certainly much cheaper to add it on to your household cover, where it will cost a few pounds a year extra. If you take out separate insurance for, say, the children's bicycles, it will be much more expensive because there will be a minimum charge to cover the administration of the policy.

Security conscious

At long last, the insurance companies are catching on to the fact that if they can persuade house owners to fit better locks they will be burgled less. Last year a massive £600 million was paid out on theft claims and that figure is rising every year.

City centres are the worst hit – and as a result the most expensive for premiums. So it is worthwhile fitting locks and burglar alarms to cut the cost of your insurance. If they prevent a burglary, they will also save you from a lot of heartache and nuisance.

The companies which offer discounts expect different things from their customers. Some will give you a 5 per cent reduction just for being in a Neighbourhood Watch scheme or fitting good locks to windows and doors. Others offer a no-claims bonus if you don't claim on your insurance policy – implying that if you have good locks you won't be burgled – or will reduce your premium if you are over fifty-five, on the basis that you stay in more in the evenings.

Others demand much more stringent security and ask you to fit particular types of burglar alarm or locks.

But remember, if you go for one of those types of discount, and leave your home forgetting to switch on the alarm, you not only leave the house unprotected against burglars, you might invalidate your insurance policy as well.

Making a claim

Tell the insurance company as soon as possible, and ask for a claims form. Once you have sent that off, wait until you hear from the insurance company before replacing or repairing the item you are claiming for. Some insurance companies will replace the item themselves rather than send you a cheque for the claim, because they can get a discount on some goods.

If your claim is a big one, the insurance company may send an inspector or loss adjuster, to assess the damage, so keep any remains from a fire, flood or breakage, as proof, until your insurers give you permission to throw them away.

If you are in any doubt as to what you should be doing, ring or write to the insurance company for advice.

Complaints

If your claim is not settled quickly, or to your satisfaction, there are several things you can do.

Start by dealing with the company at branch level, and work up to the senior management at head office.

If that does not produce results you can take your complaint to one of the following watch-dog bodies:

- The Association of British Insurers. They have no power over insurance companies but they can get your complaint re-examined at a senior executive level. The ABI is at Aldermary House, Queen Street, London EC4N 1TT (tel. 071 600 3333).
- The Insurance Ombudsman. His decision is binding on the insurance company but check, first, that the company is a member of the Insurance Ombudsman Bureau. His decision is not binding on you. If you don't accept his findings you can still take your case to court. You cannot approach him without first trying to get your complaint sorted out within the insurance company. The Insurance Ombudsman is at 135 Park Street, London SE1 9EA (tel. 071 928 7600).
- The Personal Insurance Arbitration Service. You can only approach this service through your insurance company and the decision is binding on both the insurance company and you. So you forfeit your right to take it further through a court. This service only deals with member companies and the PIAS will only deal with claims up to £250,000. The PIAS is at the Chartered Institute of Arbitrators, 24 Angel Gate, London EC1V 2RS (tel. 071 837 4483).
- If you are insured through Lloyd's, complain through your broker first, then move on to the Manager, Advisory Department, Lloyd's, 51 Lime Street, London EC3M 7HA (tel. 071 623 7100).

MOTOR INSURANCE

If you have wheels – whether on a car, a motor bike or a lorry – you have to buy insurance for it. That is the law.

Car insurance

However, when it comes to car insurance very few companies will sell you insurance that covers only the legal minimum – that type of policy would pay out if you kill or injure someone with your car, or someone who is a passenger in your car, or damage someone's property, and for nothing else. For most of us the choices are between third-party, fire and theft, and comprehensive insurance.

Third-party, fire and theft

If you cause an accident this policy covers you for injury to anyone except the driver and for any damage done by your car, if it is your fault. It also pays out if your car is stolen or burnt. It doesn't pay for any damage done to your car unless it happens while it is stolen.

Comprehensive

A comprehensive policy covers you for everything that a third-party, fire and theft policy offers, but it also pays for any damage done to your car in an accident no matter who was to blame. Also it often covers anything in your car that is stolen from it, up to a certain amount. (It is usually £100 but the exact figure will be stated on your policy.)

Cost of car insurance

How much you pay to insure your car depends on five factors:

- the type of car
- where you live and whether your car is garaged at night
- what you use your car for
- the driver's age and driving experience
- the type of cover you choose.

Car type. Cars are arranged into some twenty bands by insurance companies. The more expensive the car is to repair, the more powerful the engine and the more likely it is to be stolen or broken into, the higher will be the premium.

Some of the 'hot hatch' cars, such as the Sierra Cosworth and the XR3i, have become almost uninsurable by some companies. And those that will take the risk add up to £250 on to the premium if the driver lives in a high-risk urban area.

Although in general the bands are the same throughout all insurance companies, some cars rate higher or lower with some companies than with others. So if you think your car is in too high a class with your insurance company, check with other companies to see if it gets a lower rating there.

Where you live. You'll pay more for your insurance if you live in a city, where there is a greater chance of you having an accident or getting your car vandalized or stolen, than if you live in a quiet town or rural area. Car insurance companies are paying out more than £1 million a day on theft claims alone so it is an area coming under close scrutiny. It is no longer enough to fit a DIY car alarm – after all how many people rush out to check a car if they hear the alarm going off for the fifth time in a day.

Insurance companies have now moved on from incentives (giving you a discount for fitting an alarm) to penalties. Unless you have a proper alarm fitted by a garage, or keep your car in a garage overnight, you could be liable for the first £100 of any theft claim.

Some insurance companies will give you a discount on your premium if your fit a proper immobilizer which prevents a thief stealing your car by hot-rodding the wires. But at £350 or so to fit, it may only suit owners of expensive vehicles to go for this discount.

If you live in a city suburb check around with insurance companies in case one or two of them rate your area lower than others. City drivers may cut costs if they have, and use, a garage for their car.

Business or pleasure. If you use your car for work it will be more expensive to insure than if the car is just for home use – though driving to and from work should not raise your premium. But check with your insurance company, just in case it does.

Driver's age. Age will affect your premium if you are young or old. The under twenty-fives tend to pay more because in general they cause accidents, perhaps because they drive fast and recklessly! The over fifty-fives pay less because in general they are slower and more careful and cause fewer accidents.

There are a lot of age bands in between but the cost of insurance rises most dramatically when you are young, and falls more dramatically when you are older.

Women are usually seen as being better drivers – or at least they make fewer claims; some insurers will actually give a discount if all the drivers are female!

Type of insurance. Third-party, fire and theft policies are cheaper than fully comprehensive ones, because they offer less cover.

No-claims bonus

If you don't claim on your insurance policy you will start to build up a no-claims bonus. Year by year, as it grows, you will pay a percentage less of the quoted premium until you reach the maximum no-claims bonus of 60 or 65 per cent.

A common scale would be:

1 year with no claim	30 per cent discount
2 years with no claim	40 per cent discount
3 years with no claim	50 per cent discount
4 years with no claim	60 per cent discount.

If you make a claim you usually move back two steps, so when the no-claims bonus has built up to a 50 or 60 per cent discount, it can be cheaper to pay for minor damage yourself and keep your bonus, rather than claiming on your policy and losing the bonus.

Some policies allow you to 'protect' your no-claims bonus. You would be allowed to make up to two claims over a five-year period without losing your full bonus. Many insurers will give you this protection if you drive for a further two years without a claim after reaching the maximum discount. Others make a charge for a protected no-claims discount – usually of around 5 per cent of the premium.

Some companies give you a 65 per cent discount after four years without a claim or you can have a 60 per cent discount protected free of charge.

The other main way of cutting the cost of your premium is to take out an excess. This means that you pay the first part of every claim yourself. The excess can be £25, £50, £100, even £200, and the larger the excess the more of a discount they will give you. Young drivers often have a compulsory excess – but they can increase it to cut the premium cost.

Buying car insurance

When you pay your premium you must get a cover note to prove that the car is insured, and that should last until you get the official certificate of motor insurance. Towards the end of the year you will be sent a notice of renewal. At that stage you can either shop around for cheaper cover or send a cheque to renew with the same company. But whatever you do, do something. There are no days of grace with motor insurance. No premium means no cover.

Making a claim

If you have an accident you must tell your insurance company – even if you don't intend making a claim. You should take the name and address of the other driver involved and that of any independent witnesses, and try to get the name of the other driver's insurance company.

It is a good idea to make a detailed note of what happened as soon as possible after the accident. Insurers (and, if it comes to it, courts) are much more likely to believe your version if you have a record made immediately after the event to remind yourself.

Fill in an accident or claims form and, if you intend to claim, wait for the go-ahead from your insurance company before you have your car repaired. They will probably want a damage assessor to see it first, or ask you to take your car to a particular garage, or just require several estimates before any work is done on the car.

Write-off

If the car is not worth repairing the insurance company will offer you a lump sum for the car. If you don't think it is enough, haggle. Show them that other cars of the same make and year are going for much more in your area, by sending them adverts for similar cars from your local newspaper. When you do settle, the remains of your old car belong to the insurance company.

Knock-for-knock

If two cars involved in an accident are covered by comprehensive policies with different insurance companies you might find yourself on the wrong end of a knock-for-knock agreement. That means that each insurance company pays for the damage done to the car it insures, irrespective of who was to blame. This cuts down administration costs which helps to keep premiums down.

However, it could lose you your no-claims bonus. So if you were not to blame for the accident make sure that your bonus is safe. If there were independent witnesses, or the police are prosecuting the other driver, make sure the insurance company knows about it.

Your insurance company will not tell you as a matter of course if they are settling under a knock-for-knock agreement so write and ask them.

Complaints

The complaints procedure for car insurance is the same as that for house insurance. See p. 198.

Cutting the cost of car insurance

1. Build up a no-claims bonus.
2. Only have one named driver – preferably your partner.
3. Keep the children off your policy – let them take out their own.
4. Increase the size of the voluntary excess.
5. Don't use the car for work.
6. Don't buy a 'hot hatch'.
7. Don't drink and drive – when you get your licence back your premiums will be sky high.
8. If you are over fifty use an insurance company that offers a discount.
9. Check that you don't lose your no-claims bonus through a knock-for-knock agreement.
10. See if your union, association or club has a special scheme at low rates.

Insuring a motor cycle

Because of the high level of risk involved in owning and driving a motor bike, many insurance companies do not offer insurance at all, or will only give third-party cover.

Where you live is not so crucial when it comes to the cost of the premium – but your age and type of motor bike are vital. Premiums for youths under the age of twenty-five are much higher than for all other age groups, and the cubic capacity of the bike and the length of time the driver has held his licence will also affect the cost of the insurance.

No-claims bonuses work on the same principle as for a car, but they tend to be smaller, starting at 10 per cent and rising over three years to 20 per cent. And if you do have an accident you'll probably find your whole bonus will be lost.

All policies carry an excess clause – often the younger you are, the higher the excess.

Insuring a caravan

If you intend towing a caravan behind your car, let the insurance company know in advance. Most policies automatically give third-party cover providing the caravan is properly attached.

If you want more cover than that, perhaps insurance for the contents and the caravan itself, check your household contents policy to see if your caravan can be added to that.

Alternatively you may be able to include the caravan on a separate policy for sports equipment if you have one. Or if you belong to a caravan touring club it will probably have an insurance tie-up with one of the major companies.

Or you could go the most expensive road, and take out a specific policy yourself to cover your caravan.

Insuring your car abroad

Although your car insurance policy for this country will give you cover in EC countries to meet with their minimum legal requirements, it won't be enough if you have an accident. It does not include any cover for theft or damage to your car and may not completely cover your legal liabilities to other people.

If you want your UK level of cover extended to your holiday destination, tell your insurance company. They will extend your cover and some may offer you a 'green card'; that's the familiar term for an international certificate of motor insurance, usually for a flat fee of between £10 and £20. Some countries, such as Poland, Israel, and Morocco insist that you have a green card, or that you buy expensive local insurance. And just try claiming on that!

If you are travelling to Spain, you will need a Bail Bond. If you have an accident – even through no fault of your own – the Spanish authorities are within their rights to detain you and impound your car. To free yourself you will need a deposit against the possibility of your being found responsible for the accident or traffic violation. A Bail Bond is the answer.

Your insurance company can supply you with a Bond which the Spanish authorities will accept.

HOLIDAY INSURANCE

This is not an insurance to skimp on. If you can afford the holiday, you can afford the insurance.

If you are going on a package holiday with a reputable company it is often a good idea to take out the insurance they provide with the package – indeed, they sometimes insist that you do. It will be tailored to your type of holiday and if anything goes wrong the rep in the resort should be able to help you. If the insurance company doesn't pay up on your return, you can always try writing to the holiday company for help.

Whether you are buying the holiday insurance that goes with the package holiday or any other scheme there are certain areas you should check.

Cancellation

In case you have to cancel for a valid reason make sure the policy covers all the expenses of yourself and the people you are travelling with. If the policy only agrees to cover you and your spouse this is no good if you are travelling with a friend.

Valid reasons should include illness or death of yourself, a close relative or business partner or being called for jury service. The more expensive policies also cover you for redundancy, a burglary or some other sort of disaster at home or the police asking you not to go away because of a burglary at home or work.

Medical expenses

Bills for hospital treatment abroad, particularly in America, can be huge, so make sure the policy offers at least £250,000 of cover for Europe, £1–2 million for elsewhere. It should also cover doctors' bills, and the cost of any medicines you have to buy – though there's usually an excess here, meaning that you have to pay the first £5 or £10 yourself.

Some, more expensive, policies also pay to fly you home by air ambulance if necessary and will pay for any extra accommodation expenses of a friend or relative who stays on with you.

If you are travelling to an EC country you do get some sort of reciprocal health cover from their health service. You have to sign form E111 and get it stamped by your local Department of Social Security office before you go, and it can be a nuisance trying to use it. With some countries you have to pay up and claim when you get home, with others you can only use certain doctors or hospitals. It is much easier to take out proper holiday insurance.

Watch out

1 Hazardous activities such as water skiing, para-gliding, pot-holing and so on are often excluded from the policies, so you won't be covered if you injure yourself through doing a particularly dangerous sport. On winter-sport holidays, skiing, tobogganing and so on are included, which is why skiing insurance is more expensive than summer holiday insurance.
2 Pregnancy, and any complication arising from it, are not automatically covered. So if you think there is a chance, look for an insurance policy that covers you.
3 Pre-existing illnesses. They either won't be covered or you will have to get a doctor's letter saying you are fit to travel. If you do have asthma or whatever, take your medication with you.
4 Drink and drugs – if you were under the influence and injured yourself, the policy probably won't pay up.
5 Private medical care or dental treatment which you have in this country to clear up something you picked up abroad will not be paid for.
6 Age – the upper age limit for travellers varies from policy to policy so if you are over seventy, look around and find one that will cover you but doesn't charge extra for oldsters.

Luggage and money

This part of the insurance will cover anything that is lost or stolen from you – providing you were not unduly careless. If you leave your camera carelessly on your towel while you go in swimming, and it is stolen, you won't be able to claim. All insurance companies expect you to take reasonable care of your belongings.

Your traveller's cheques, plastic cards and up to around £200 of cash will also be covered if they are lost or stolen.

If you have a household contents policy, it may also give you some cover. But personal belongings and money insurance comes as part of the overall holiday package so you might as well use it. It will give you insurance for up to around £1,000 for lost or stolen belongings but there is usually a limit of £250 on any one item. So if you have an expensive camera it won't be covered. Add it, all risks, on to your household policy. In order to claim on your holiday insurance, you must report anything that's lost or stolen to the local police and your holiday rep – if there is one – and remember to get written proof that you have reported it to them.

Personal liability

If you injure or damage someone or their property while you are on holiday most policies will pay out. You should make sure the limit is at least £500,000 in Europe, £2 million elsewhere, and that the policy doesn't exclude claims made in courts outside the UK. It is most likely that the claim will be made in the courts of the country in which the incident took place.

Claims as a result of your driving a car, motor bike, plane, or boat are excluded, as are claims made on you by a member of your family.

Personal accident

Macabre as it may seem, the larger the number of limbs and eyes that you lose, the more money you get. But it isn't much – ranging from £5,000 for the loss of a leg to £25,000 if you die. Sometimes it is paid as a lump sum, sometimes as an amount per month.

If you are the only breadwinner in the family you should take out an additional form of life insurance. (For further details on life insurance see Chapter 6.)

The insurance policy won't pay out if you are taking part in a hazardous sport or are drunk or on drugs.

> **Tip**
> If you pay for your holiday, your holiday deposit or your travel arrangements with your credit card, you automatically get personal accident insurance for nothing.

Additional clauses

Most holiday insurance policies nowadays also add on all or some of the following clauses:

Delayed departure. If the plane or train is more than twelve hours late you will get a lump sum, usually up to £100, for the delay, and an additional sum for every subsequent twelve-hour period until the plane goes. There is usually a maximum sum and, as often as not, you'll find that your plane takes off eleven and a half hours late.

Missed departure. If you miss your flight because of failure of public transport due to weather, a strike (if there is no advance warning) or mechanical breakdown (not of your own car), the policy will pay your extra expenses up to a set limit.

Delayed luggage. If you get there and your suitcase doesn't you should get something, usually £75–100 if you have to wait over twelve hours. Keep receipts for any toothbrush, underwear or bikinis you buy.

Hospital benefit. If you are in hospital some policies pay your extra expenses, such as phone calls up to a limit.

Tour operator failure. Handy if the company goes bust and isn't a member of the travel agents' and tour operators' trade associations, ABTA or ATOL (it is worth checking that they are members before you book).

> **Tip**
> Pay for your holiday with your credit card (the voucher should be made out to the holiday company, not the travel agent) and you are covered anyway. No matter whether or not the company is a member of ABTA or ATOL, you get your money back if it gets into financial difficulty before you go and your holiday is cancelled, and if it happens while you are on holiday your return airfare will be guaranteed.

Regular travellers

If you travel abroad a lot on business, or take a large number of holidays, it would be cheaper to buy a policy where there is an annual fee irrespective of the number of trips you take.

> **Tip**
> It is not always easy to get insurance companies to accept holiday claims – some just ignore the first letter they receive from holiday-makers. So if your claim is a valid one, keep pestering them until they pay up.

PERMANENT HEALTH INSURANCE

If you are an employee, and you are off work through sickness or injury you should be entitled to some state benefit.

State benefit

If you are sick or injured you will usually be entitled to Statutory Sick Pay from your employer for up to twenty-eight weeks in any period of incapacity. At the moment, SSP is either £45.30 or £52.50 a week, depending on how much you earn. Those earning over £190 a week get the higher rate; those earning between £54 and £189.99 get the lower rate.

If you cannot get SSP from your employer for any reason, or you are self-employed or unemployed, you should get Sickness Benefit. It amounts to £41.20 a week if you are under retirement pension age (sixty-five for a man, sixty for a woman), £51.95 if you are over retirement age, and it is paid for twenty-eight weeks.

At the end of the twenty-eight weeks, if you are still unable to work, you can claim Invalidity Benefit of £54.15 a week. You may also qualify for Invalidity Allowance. This depends on your age:

under 40	£11.55
40–49	£7.20
50–59 men 50–54 women	£3.60

And a further additional benefit depending on how much you earn. If you can't claim Invalidity Benefit because you have not paid enough NI contributions you will instead get Severe Disablement Allowance of £32.55 a week, and an allowance for your age worked out on the scale detailed above.

PHI

If the state benefits are not enough then you must fill the gap yourself. Work out how much you will need per month to live, if you are unable to work, and take out an insurance policy for this to be paid to you. It is called permanent health insurance.

Permanent health insurance is in fact permanent ill-health insurance. It is a type of cover which pays up if you lose your income because of illness, injury or disability. What you receive from the insurance company bears no relation to your level of injury – you decide in advance how much you would need and pay a premium related to that sum.

If you are an employee. Check with your employer how long you would continue to be paid your basic salary – your PHI could be tailor-made to start the following week.

If you are self-employed. Your problem could be twofold because you not only lose your income but you may have to pay someone to do your job for you.

On top of that you may need to employ a nanny or housekeeper to run the house, and you will have increased medical and hospital expenses which will need to be taken care of.

What to do

1 Calculate how much you will need per month to live, if you are unable to work. This should include the cost of any permanent nanny or housekeeper you might need.
2 Decide when you would like the money to start coming in – after eight weeks, thirteen weeks, twenty-six weeks or fifty-two weeks. The longer you defer the payment the cheaper is the premium.
3 Work out how long you would want the money to be paid: until age fifty, retirement age, or death.

The longer the money could potentially be paid, the higher the premium. The older you are when you take out a PHI policy the more expensive it will be (up to age fifty-five, which is the upper age limit for taking out most policies).

PRIVATE MEDICAL INSURANCE

If you want a private hospital room or don't fancy waiting for an operation, private medical insurance is probably what you need.

You pay an annual or monthly premium and in return you will get:

- consultation with a specialist
- hospital accommodation and nursing
- any operation or other treatment you need
- any drugs you need
- limited home nursing
- daily cash to cover small expenses from some companies.

To qualify you must initially be referred by your GP, and there are upper limits, in most schemes, on the total pay out.

Pregnancy, cosmetic surgery, pre-existing conditions and health screening are generally excluded, so often are psychiatrics and geriatrics.

Aids. Because Aids is still a fatal disease it is treated in different ways by the medical insurance companies. Some totally exclude Aids and HIV-related diseases, others will only cover it if it has been contracted a certain period after you joined the scheme.

The cost

Private medical insurance is not cheap, and it operates on three scales. The most expensive is for treatment in a London NHS teaching hospital

Basic BUPA Care Subscriptions (monthly)

Age groups		*Scale A*	*Scale B*	*Scale C*
18–24	Single	£58.10	£33.25	£27.78
	Married	£106.51	£60.94	£50.94
	Family	£133.14	£76.22	£63.68
25–9	Single	£58.31	£33.37	£27.90
	Married	£111.69	£63.92	£53.42
	Family	£139.63	£79.92	£66.76
30–34	Single	£60.18	£34.43	£28.77
	Married	£115.39	£66.09	£55.22
	Family	£144.22	£82.61	£69.00
35–9	Single	£62.85	£36.30	£30.33
	Married	£120.48	£69.65	£58.18
	Family	£150.57	£87.04	£72.72
40–44	Single	£64.96	£37.52	£31.34
	Married	£124.46	£71.96	£60.12
	Family	£155.54	£89.93	£75.14
45–9	Single	£68.96	£39.51	£32.99
	Married	£132.09	£75.72	£63.25
	Family	£165.09	£94.62	£79.07
50–54	Single	£77.64	£44.71	£37.29
	Married	£148.92	£85.82	£71.53
	Family	£186.17	£107.27	£89.44
55–9	Single	£90.12	£51.59	£43.02
	Married	£172.71	£98.91	£82.42
	Family	£215.90	£123.63	£103.06
60–64	Single	£101.75	£58.35	£48.65
	Married	£194.88	£111.83	£93.20
	Standard rates on request.			

or private hospital, then come the higher-cost independent hospitals outside London and finally the large provincial NHS teaching hospitals and the cheaper independents. You can choose which scale you want or can afford, and pay the appropriate premium.

These charges will vary from company to company, and you can often reduce the premiums considerably if you opt for one of the companies which requires you to try for an NHS operation first. If the waiting-list is longer than six weeks then they will pay for you to have it privately (without having to wait the six weeks first), if it is shorter than six weeks then you get it done on the NHS, and receive a daily allowance from the medical insurer to cover your incidental expenses.

Some companies also operate an excess to reduce costs.

The insurance policies also operate upper limits on consultants' fees, above which they will not pay, but most consultants, surgeons and anaesthetists know these scales and they generally keep within them.

Likely hospital and specialist charges

Hernia	£600–£1,200
Hip replacement	£3,800
Varicose veins	£1,200
Hysterectomy	£3,600
Tonsillectomy	£1,000

Cut price

Because of the high cost of this type of insurance, the companies involved offer plans that cut the cost, sometimes at the expense of reducing the cover. You can cut the cost in the following ways:

- Join through your company, union or association, they have often agreed a discount with a particular insurance company.
- Join young. The older you are when you join, the more expensive it is.
- Pay by credit card or direct debit; it is cheaper.
- Don't claim unless it is necessary. With some schemes you can get up to a 40 per cent no-claims bonus after five years.
- Agree to try and get the operation on the NHS first. If you have to wait longer than six weeks you can have the operation privately – but your premiums will be much lower because you agreed to go the NHS route first. You will also get a cash pay-out of anything from

£15 to £40 for every day you are in hospital if the op. is done on the NHS.

Tax relief for pensioners' premiums

The older you are, the higher the premiums for private medical insurance rise – simply because you are getting older and therefore more likely to need hospital or specialist treatment. Indeed, many private medical insurance schemes won't take you on as a new customer once you reach sixty. And even if they do, for many pensioners the premiums are just too high.

But two recent moves will make a difference:

- government help
- premium help.

Government help

The government is keen to encourage more people to pay for their own medical treatment so that NHS queues for operations and treatment will be shorter and the cost to the NHS lower. So tax relief on pensioners' premiums has been introduced. It works a bit like MIRAS does on mortgages.

Pensioners pay their premiums net of 25 per cent tax to the medical insurance companies. So, in effect, premiums become a quarter cheaper. Pensioners who are higher-rate tax payers – that is, they pay 40 per cent tax on some of their income – will qualify for higher-rate tax rebates. If that's you, tell your tax inspector and your tax code will be adjusted accordingly.

The tax relief is given to whoever pays the premiums, providing the policyholder is over sixty. So a generous son or daughter or friend could get tax relief at their top rate on a parent's policy. The only exception is that a company still paying the premiums on a pensioner's policy would not qualify for any tax relief.

Premium help

To reduce premiums still further, budget or low-cost schemes have been introduced to help older people. You can now join medical health insurance schemes up to age seventy-four, but of course it will be an expensive business to pay for full cover because the chance of illness, and particularly serious illness, is higher at this age.

Much of the cover offered to older people through the low-cost schemes assumes that they will be prepared to have the operation on the NHS, providing they don't have to wait too long for it. And out-patient treatment is only paid if this relates directly to in-patient treatment.

Opting for an excess of, say £100, will also reduce your subscription. For example, a seventy-one-year-old man would pay

without excess:	with £100 excess:
£43.30 per month	£41.10 per month

How to claim

If you are in one of the scheme's private hospitals, you will not see any bills at all. Otherwise try to get the hospital to deal with the insurance company directly. If they won't, send in your receipts with the claim. If paying the bill and then claiming could put you into the red, tell the insurance company before you have the operation and it may be able to speed up payment.

> **Tip**
> Before going to see a specialist or having any private treatment, check with your medical health insurance company that you are covered, and that your consultant's fees fall within the scale you have chosen. Otherwise you could be in for a nasty financial shock.

INSURING AGAINST LEGAL EXPENSES

Legal insurance is expensive mainly because once people have it, they use it. No one wants to have a fire or a burglary, but a court battle at someone else's expense is a challenge worth taking up.

You can buy legal expenses insurance in two ways:

- an add-on to your car or household insurance
- a separate legal protection policy.

The policies will cover the solicitors' and barristers' fees and the other side's legal expenses if you lose the case.

Legal expenses insurance is often offered as an inducement to other financial services nowadays. If you take out a loan, or buy house insurance, you might be offered a free legal service for nothing. This may not actually cover your legal costs if you go to court, but it will give you advice on whether or not to proceed with a case, or what to do if you are in legal difficulties.

Add-ons

As an add-on, legal insurance is relatively cheap, usually only a few pounds a year. If you add it on to your car insurance policy, you can claim for 'uninsured loss recovery'. That means if the accident was not your fault, you could sue to recover your excess (if you have had to pay it), the cost of hiring a car while yours was off the road or loss of earnings.

If you add it on to your household insurance, you will be able to fight consumer disputes, civil actions relating to houses, employment disputes, personal injury or defending minor motoring offences.

They will provide you with cover of between £10,000 and £25,000 a year.

Separate legal insurance

Family policies are much more expensive if bought in their own right. You can pay anything from £50 to £150 a year depending on the level of cover you want.

The policies will cover you for most disputes, providing the company thinks you have a reasonable chance of winning, otherwise it won't let you go ahead.

But there are exclusions:

- divorce
- disputes relating to major building works
- landlord and tenant disputes if you live in a block of flats.

And you often have to pay the first £30 of any legal fees yourself.

The better policies have a twenty-four-hour telephone hot-line which allows you to phone for advice at any time. (Check that it is not just an answering service out of office hours.)

The major insurance companies in this business are DAS, Allianz, Hambro Legal Protection and Legal Protection Group.

REDUNDANCY INSURANCE

The insurance industry is against redundancy insurance. Because of the random nature of job losses it is difficult to assess statistically, so few offer it as a policy.

What you can get is redundancy cover on any loan you take out, including mortgages. Building societies, banks, credit cards, HP accounts, can all be covered. The insurance doesn't wipe out the debt, it guarantees the payments for twelve to twenty-four months. But that could give you just the breathing space you need to get back on your feet.

The cost varies enormously depending on what you are covering. But a mortgage of £200 a month would set you back £9 to £10 a month for redundancy insurance, while an in-store credit card would be a mere £1 a month to cover a similar repayment.

OTHER INSURANCE

There are other types of insurance which, although not bought automatically by a large percentage of the population, are none the less fairly standard. Pluvious policies will cover your 'event' – whether it is a wedding, a church fête or a charity cricket match – against being ruined by rain. Pets' insurance plans will cover your animals against the cost of vets' fees and medicines if they are sick.

You can buy specific insurance for your musical instruments, any valuable collection you have, small boats or computers.

Generally, pound for pound, these types of insurance are more expensive and you have to search around to find some of them.

However, if you try to go for plans that are already set up by insurance companies rather than expecting your company to set up a package specially for you, then you should get as good a deal as there is going. If you want to insure something associated with your hobby, check clubs or specialist magazines for details of specially set-up schemes.

BUYING INSURANCE

When it comes to taking out a policy you can do it in one of two ways:

- directly with the insurance company
- through an intermediary – a broker.

It should cost the same whichever route you choose.

If you deal with a broker he should, in theory, be able to get you the best value for money. In practice most intermediaries have a short-list of companies they tend to use most of the time and commission rates vary from insurance company to insurance company so you could unwittingly be pushed towards the one paying the highest commission rather than the one which would be best for you.

But an intermediary will do all the paperwork for you and help you with any claims you make.

Use an insurance broker who is a member of the British Insurance and Investment Brokers' Association.

19 Children and Money

Investing for babies – the options · Investing for children · Children and income tax · How to afford school fees

See also
- *Chapter 4, Buying and Selling Shares*

Whatever else babies are born with – joyful smiles, an ability to sleep through the night – they come into the world with two financial advantages:

- a right to Child Benefit for at least sixteen years
- their own personal tax-free allowance.

Collecting the first and making use of the second will set your child on the right money path for life.

Child Benefit

Once your baby is born you are entitled to claim Child Benefit for him or her. The payment (£9.65 a week, or £7.80 for every child after the first) is not means tested – you receive it no matter how rich or poor you are. You can pick up the cash at your local post office every month or you can have it paid directly into a bank or building society account. But if you are a one-parent family you can choose to be paid weekly.

To claim the benefit, fill in the leaflet in the Department of Social Security booklet FB8, 'Babies and Benefits', and send it off along with the child's birth certificate.

Child Benefit is paid monthly in arrears until children are sixteen, or until they are nineteen if they continue in full-time education.

If the parents of the child live together, then the mother should claim the benefit (it doesn't matter whether or not the parents are married) but if the parents split up then either partner can claim, depending on which one the child lives with.

Tax and children

All children are born with their own personal tax allowance. The problem is they are not born with any money so they can't make use of the allowance until someone gives them some.

The allowance they have is the personal tax-free allowance. If they are successful child models, or have a lucrative paper round at a very early age, they can earn up to the personal tax allowance level without having to pay tax on the money.

Or if they are given money and it is put in a savings account or in shares or unit trusts or whatever in the child's name the interest or income on the money can build up tax-free . . . unless the money is given by a parent.

If the money is given by a parent

When the interest is more than £100 a year, the Inland Revenue will assume the money is still the parents' and tax it anyway. That rule is designed to stop parents salting cash into their child's account to avoid paying tax on the interest. Unless the child becomes so wealthy that she breaches the personal tax-free allowance, the Inland Revenue does not expect her to fill in a tax form. However, it is the duty of the parents to inform the local tax office if they feel that tax is due.

INVESTING FOR BABIES

Many parents, grandparents, godparents and friends want to invest for newborn babies – either as a christening gift or as a present for their future. For once, the first criterion is not safety. Whether it is a one-off present or an annual payment, most people want the capital to grow over the years so that it will turn into a nice lump sum by the child's eighteenth birthday.

Building society, bank and National Savings accounts are the avenues you could take if you want a completely risk-free investment. Your capital will be safe and have interest added for eighteen years. But that's all.

For most people an investment linked to the stock market is a better option. Because you are thinking of investing over the very long term, up to eighteen years or more, the risk is low because over that length of time shares should do well.

There are several ways you can invest:

- You could buy shares directly (for details on how to buy shares, see Chapter 4). If you do that, you could probably lock away an investment in some of Britain's top companies and over the eighteen years see the share price and dividend income steadily grow. If you buy shares in smaller companies or anything risky, you will have to monitor the shares periodically and buy and sell to take advantage of price rises. Children under eighteen cannot sell shares, so the parent would have to be named as a trustee.
- You could buy unit or investment trusts (see p. 21). An investment in either of these types of funds should grow steadily, particularly if you choose a fund that has been set up to provide capital growth. You could buy direct or invest in a PEP (see p. 46). Some unit trust companies have schemes specifically designed for investments for children. The giver agrees to pay a regular monthly or annual premium and the unit trust group provides all the necessary forms so that tax can be claimed back from dividends. In England and Wales, children have to be over fourteen to hold units in their own name, and over eighteen to sell the units. In Scotland they have to be over eighteen to hold or sell units.
- You could buy a unit-linked endowment policy (see p. 56). This is a form of investment which has to run for at least ten years. You pay a lump sum, or make annual or monthly contributions, and the money is invested, usually, in shares and gilts.

CHILD ACCOUNTS

Once children reach the age of seven and can sign their own name, it is a good idea to encourage them to open their own bank, building society or National Savings account. Good saving habits start young and once they get the idea of using a bank or building society to save up for something that they want, and see the interest being added regularly, they should be able to continue the habit. And then of course there are the free gifts.

A child can, of course, open an account at a younger age but one of the parents would have to act as trustee. Many young children like to have their own account, and get the free bag of goodies that often goes with it.

Some of these accounts designed specifically for children offer a poor rate of interest. That's because they attract just the sort of customer that the bank or building society doesn't really want. A person who has not got a large balance, but comes in personally to pay in or withdraw small sums of money.

None the less, an account like this is invaluable to the child. It teaches them how a financial institution works, gets them into the habit of saving and watching the funds grow and prevents them having an adult fear of using a bank or building society properly.

If the nest egg starts to build up to more than, say, £100 then you should perhaps think about a second account; one that pays more interest in return for perhaps needing notice of withdrawals.

Tip
Remember to sign a form ensuring that the interest is paid gross – or you'll lose tax unnecessarily.

Banks and building societies work on the principle that once you open an account, you keep it open for life. Junior who is paying in £1 a week or even £1 a year now is costing the bank or building society money to keep such a small account running, but in the years to come he will probably want an overdraft, a mortgage, a personal loan, maybe even help to start up a business. That is when the bank or building society's investment will pay off.

So to attract them young all banks and several of the building societies offer packages of free gifts to children who open accounts. The free offers change when the children become teenagers or students but for the tiny tots the free gifts tend to be along the lines of piggy banks and tee-shirts.

National Savings

Any building society, bank or National Savings account that you open in the child's name can only *accept* cash for the first seven years. You won't be able to get any of the money back out of the account until the child is old enough to sign his or her own name in 'joined up' writing. In theory, the only exception to that is if you are emigrating or if you can prove actual hardship for the children that would be alleviated by their having access to their own money. In practice, if you write to the bank or building society, or to the Department of National Savings and say you need the money for the child, perhaps to pay for school fees or a new school uniform, you would probably get it. Alternatively, make sure a parent is a trustee of the account.

The cost of private education (per term)		
	Boarding	*Day*
Boys' senior school	£1,900–£3,600	£900–£2,500
Girls' senior school	£1,900–£3,400	£900–£1,900
Preparatory	£1,300–£2,500	£600–£1,650
Pre-preparatory		£300–£700

SCHOOL FEES

Rising incomes and worries over education cuts are encouraging more parents than ever before to look at the option of educating their children independently.

Few can afford to pay the huge fees out of annual income; for most parents the financial burden of funding their children's education is enormous.

There are four main ways of funding school fees:

- capital payment in advance
- insurance policies in advance
- borrowing the money now and paying it back later
- sharing the burden with other members of the family.

The earlier you start to save for school fees, the easier it will be to pay them when the time comes.

Capital payment in advance

This type of scheme suits parents who have a lump sum set aside to pay their children's fees in the future. If you use a scheme specially set up by one of the large school fee specialists it will be tailored to suit your particular needs. The money will start to be made available in the first term the child goes to school, if that is what you want, or it will be left in the fund continuing to grow until your child reaches the age when you need the money for fees.

Some plans are guaranteed, and you know at the outset how much they will pay out per term or per year, but this 'guarantee' could cost you by lowering the return.

Others offer non-guaranteed options, which means they grow as fast as the investments. The cash will be linked to shares, gilts, unit trusts, building society or National Savings accounts and the spread of investments will depend on the investor's rate of tax, how much risk he is prepared to take and how long the money is invested.

If you prefer, you could place your money in a capital growth unit trust not in any way connected to school fees, and cash in units as you need the money. But you could be caught short if you need to cash in the units at a time when the stock market, and therefore unit trust prices, are falling. To avoid this risk parents can use educational trusts to guarantee their child's fees. They pay a lump sum to an insurance company, and the money is used to buy an annuity which will guarantee a certain level of fees at a given date. These trusts have charitable status and are tax efficient to high-rate taxpayers but the returns are low, because they are guaranteed, and they are particularly inflexible. The charitable school fees educational trusts are run by Save & Prosper, Royal Life, School Fees Insurance Agency and Equitable Life.

Advantages. Ideal for parents with a lump sum.

Disadvantages. No good for parents who decide on independent schooling late or who have no savings in the early years of marriage.

Insurance policies in advance

This is still the main one, and is well suited to parents wanting to save regularly now to pay fees in the future.

Parents pay a monthly or annual sum into an endowment policy and the sum builds up and is increased annually by a bonus that the life insurance company adds on. A further bonus is added at the end. If the parent dies during the term of the endowment policy, the lump sum assured is paid out by the life insurance company so the fees will be covered.

You can either take out separate policies timed to mature in each of the years fees are needed or have a large number of 'flexible' policies which you mature when you need them. Many parents top up these schemes using a capital sum or part of their income to fund any short fall or pay the fees direct in earlier years.

Advantages. A good way of saving and building on your savings for future fees.

Disadvantages. Endowment plans have to run for at least five years and early surrender is heavily penalized, so you have to start early. Not so good for prep school fees unless you start the policies before the child is born.

Borrow now, pay later

If you decide to educate your child independently, and want to start her at the new school next term, advance planning won't come into it. You need the money now.

The best way for parents with this problem to get the cash is to borrow it using their house as security. You can borrow up to 70 per cent of the value of the house, less any existing mortgage, and protect the loan using a with-profits endowment policy. Interest is only paid on the money you have withdrawn.

Advantages. Good for parents who did no advance planning.

Disadvantages. Parents will be paying for the child's education long after she has grown up, and the repayments will be large as these loans do not qualify for the tax relief that mortgages do. This method should be used only as a last resort because you give the lender a second charge on your home and this is a very serious step to take. If you fail to make the payments, you could lose your home.

Sharing the cost

In many families it is not just the parents who contribute to school fees. Increasingly grandparents offer to do what they can to help.

If the parents are planning early and getting schemes set up long before the child goes to school, the money that the grandparent gives can be used as the premium payment for a regular savings plan or endowment policy in the child's name.

Since children can no longer claim back the tax that the grandparents have already paid through covenants, new types of tax-efficient schemes are being set up. The grandparent takes out an annuity which pays an income, gross, to the child. And the child can keep the money, tax-free, up to the level of the personal tax-free allowance.

A lump sum of £20,000 would currently pay out £2,600 a year. If there is still time, the annuity can fund endowment premiums rather than the school fees themselves, and the endowment lump-sum pay-out will fund the fees.

Parents cannot buy the annuity for this purpose because the income will be deemed to be theirs and taxed accordingly.

Advantages. Releases parents from the whole burden of school fees.

Disadvantages. Not everyone has wealthy or generous relatives.

Care and maintenance trusts

If a generous grandparent wants to give a child a regular income, perhaps to fund the school fees bill, the best way to do it is through a care and maintenance trust. The grandparent (or whoever has made the payment) sets up the trust with a lump sum, and providing the grandparent then lives for seven years, no inheritance tax will have to be paid on the money.

Parents cannot give money to their children in this way.

Once the trust is set up, income from the investments builds up within the trust. The trustees have to make provision to pay tax on the income but it doesn't actually have to be paid because the child can claim this tax back – up to the personal allowance every year – if the money is being used for 'maintenance, education or other benefit'. So that includes school fees.

It is up to the trustees, who can be the parents, to decide when the income of the fund is paid out. But once the child reaches the age of twenty-five the income must be paid annually.

The lump sum that was used to set up the trust can be passed on to anyone, not just the child who benefited from the income. So, if the grandparents set up a care and maintenance trust for the grandchildren, the capital could still be left to the parents.

If you are planning on using a care and maintenance trust you will need the help of a solicitor or accountant.

Warning
A child only has one personal allowance. So all income from interest on National Savings accounts, annuities, care and maintenance trusts and whatever will be added together. Once it reaches the level of the personal allowance the child will be liable to tax in the same way that any other single person is.

Assisted Places

Even if you cannot afford the fees, your child could still attend a public school through the government-backed Assisted Places scheme. At the moment some 27,000 pupils benefit in this way.

To qualify academically, a child only has to be as good as the other children in the class. That means sitting the same entrance exam and qualifying in the normal way.

The scheme is a means-tested benefit and the fees will be paid in full for the child if parents earn less than £8,714 a year after allowances have been deducted for other dependent children. Assistance is then given on a sliding scale up to a cut-off point, when the relevant parental income reaches £21,000. Some boarding-schools will also help with the cost of living in.

Further details of this scheme are available from the Department of Education and Science, Room 3/65 Elizabeth House, York Road, London SE1 7PH; the Welsh Office Education Department, Crown Offices, Cathays Park, Cardiff CF1 3NQ; the Scottish Education Department, Room 4/25 New St Andrew's House, St James Centre, Edinburgh EH1 3SY.

USEFUL PAMPHLETS AND NAMES AND ADDRESSES

Child Benefit
DSS leaflet CH1, 'Child Benefit'
DSS leaflet FB8, 'Babies and Benefit'
School fees
Independent Schools Information Service (ISIS), 56 Buckingham Gate, London SW1E 6AG (tel. 071 630 8793).

20 Teenagers – Money, College and First Job

Savings for teenagers · Earning pocket money – how much should you get? · Students and their money · DSS benefits · Getting a job · Unemployed school leavers

See also

- *Chapter 7, Understanding Income Tax*
- *Chapter 8, Tax and Your Job*
- *Chapter 19, Children and Money*

TEENAGERS AND THEIR MONEY

By the time children get into their teens, their attitude towards money changes. Saving no longer has the same appeal. Under-twelves often save just for the sake of it – they actually like to see the cash piling up.

Teenagers have too many temptations for that to hold any appeal. If they save at all, it is because they are saving for something specific – a mountain bike, the latest Madonna CD or some new clothes.

They get their money from three sources:

- parents and relatives giving money presents
- pocket money
- part-time work.

Savings accounts

At this age, where they put their money begins to take on some importance. If they have got savings which are beginning to build up then the cash should be moved into a savings account paying a good rate of interest.

Teenagers should open a bank or building society account if they don't already have one. Many people in their twenties and thirties are still a little nervous of banks and building societies, and don't know quite how they work. Getting used to them as a teenager will prevent that happening – and it also makes sure that the children, when they grow up, have a healthy and positive feel for these institutions. Banks and building societies make their profits out of lending money, so no one should ever feel embarrassed asking for a loan, whatever the attitude of the branch manager.

Remember, teenagers should sign the form so that interest is paid gross.

The banks are just as keen to have teenagers open accounts with them as they were to attract them as youngsters. The range of goodies changes – from piggy banks to calculators and cash – but the idea of the bribe is the same.

Children can open accounts with as many banks as they like – providing they have the £1 or so needed as an opening balance. But the money in the accounts is theirs. If they want to withdraw it all to buy a stereo system or gamble it on fruit machines, that is their right. Even if the bank manager knows this is happening, it is against the law for him to tell the parents about it.

Once they reach fourteen, youngsters can be offered plastic cards for withdrawing money from 'hole-in-the-wall' machines. That gives them access to between £100 and £250 a day depending on the bank or building society (if they have that sort of cash in their account) but there are no facilities for overdrawing.

Once they get to sixteen, if they have left school either to get a job or go for further education, then they might get a cheque book and cheque card. Some cheque cards are also credit cards but the holder won't be able to run up any debts on it until coming of age at eighteen, when they can be held legally responsible for any debts run up on the card.

Pocket money

Once children become teenagers, pocket money takes on a whole new meaning. When they were youngsters, it tended to be spent on sweets or little knick-knacks from the shops, or saved.

Now it can be the main source of income for some kids, so they will expect a decent sum – particularly if the parents also expect them to buy some of their own clothes, or save up for a mountain bike.

Pocket money averages

Year	*5–7*	*8–10*	*11–13*	*14–16*	*Boys*	*Girls*	*Average*
1991	64p	139p	216p	289p	170p	167p	169p
1992	70p	143p	223p	333p	190p	174p	182p

Source: Wall's Pocket Money Monitor by Gallup

On average, fourteen-to-sixteen-year-olds get around £2 to £3 a week from their parents, with boys doing better than girls, and Scottish teenagers getting more than their English counterparts.

WORKING CHILDREN

Over half the children in this country have a part-time job by the time they reach the age of sixteen. Mostly it is a paper round or regular babysitting, but a lucky few have more glamorous work on the stage or making TV commercials or films.

None of them make much money out of their jobs.

Average wage rates

Paper rounds. Teenagers should expect to get £1 to £2 a morning with more on a Sunday and often a weekly bonus as well.

Babysitting. Teenagers should expect to get £1 to £2 an hour (more in London!), and a lift home at the end of the evening.

Acting. A part in a play or a pantomime whether it is in a provincial theatre or in the West End will pay upwards of £13 a performance.

Commercials. Children and teenagers will get around £50–£100 a day – with a further £30 going to their parent or chaperone. Babies get a mere £25 a day, with around £30 going to the parent or chaperone. But rates for adverts are more negotiable – and depend on how much the company wants the baby or child.

Television. Children and teenagers with a proper speaking part in a television programme will get between £35 and £100 a day; as an extra they will get around £24 a day. A parent or chaperone would get a further £30, or a qualified matron £35.

The reason child actors and models are paid so little is quite well founded. It is not just that there is a huge supply of kids chasing too few parts. Wages are kept deliberately low to discourage greedy parents from forcing unwilling children into work they don't want. If there is little cash about the child will only do it for fun.

Rules

We have come a long way since the days when children worked sixteen-hour shifts in factories. There are now proper rules governing working conditions for children.

Generally, the younger the child is the less he or she can work and the longer the work breaks must be.

So a babe in arms can be in the studio or theatre for three hours a day, between 9.30 a.m. and 4 p.m., and only work continuously for twenty minutes at a time.

A fifteen-year-old, by contrast, can be in a studio or theatre for eight hours between 9 a.m. and 7 p.m. and work continuously for forty-five minutes.

There is also a strict limit on the number of days a year a child can work and a child needs a special local education authority licence in most cases.

Some child-actor families have been known to move house to come into the jurisdiction of a slacker education authority!

Tip

If your child does a paper round using a bicycle, she won't be covered by your household insurance policy if she injures herself or causes an accident. So check that the newsagent has a policy to cover the paper boys and girls. Most do.

COLLECTIONS

If your child has been a mad collector of something when he was younger – be it stamps, comics, or Matchbox cars – it may be worth some money now. Or it may gain in value in years to come. It is probably worth saving if you have room in your loft or cellar.

As a rule of thumb the things which will be valuable in the future will be rare, complete sets and in good condition. And if it is a set of toys make sure that they are all in the boxes they came in – a forlorn hope!

COMING OF AGE

At eighteen, the world is your oyster. You can have your first legal drink in a pub, vote in elections and you might already be able to drive a car.

In financial terms it is a crucial age – because at eighteen you are deemed to be old enough to be responsible for running up your own debts. You can buy and sell shares, unit trusts and investment trusts in your own name, take out HP agreements, run up a credit card bill, sign up for a mortgage. And, of course, be responsible for the bills that come in.

By the time they reach this age, most teenagers will have some idea about handling money and probably already have a bank or building society account. What they do next, in money terms, depends on what sort of future lies ahead for them. Leaving school usually means taking one of three roads:

- going on to university or college
- getting a job
- becoming unemployed.

STUDENTS AND THEIR MONEY

If you are going on to university or college, and you qualify for a grant, the beginning of your first term will probably be the first time you have to handle a large sum of money.

It will seem a lot when you receive the cheque but in terms of pounds per week it is not much. So it is important that you learn to budget.

The first step should be to open a bank account near your college or university. These are the branches that the banks assume students will use and are geared up for that clientele, with such advantages as student advisers or help for the unlucky student whose grant comes through late.

All banks offer packages to students to tempt them to open an account. Look at what is on offer from them all – including some of the smaller banks such as the Yorkshire and the Co-op – and choose the scheme which suits you best. It may offer a cash incentive, free overdraft facility or interest on the money in the account – only you know which one will be the most useful for you.

Income	Winter	Spring	Summer
Grant			
Parental contribution			
Vacation earnings			
Anything else			
Total			
Outgoings			
Accommodation			
Food			
Gas and electricity			
Fares			
Clothes			
Books			
Entertainment			
Total			

Student budget planner

Tip
Make sure that the bank adds the cash that you are due. Banks do not have an automatic blip on their computer to tell that a cheque is a grant cheque, so when you pay it in make it clear to the cashier. Otherwise you won't get the incentive money credited that is yours by right.

You must work out a weekly budget for yourself. Use the student budget planner to work out how much you will have coming in every term, and how much you think you will have going out. If there is more going out than coming in you will have to cut down, or find an alternative source of income. If there is more coming in than going out, you will know how much you have as spending money every week. But remember to allow some cash for emergencies.

The grant

Your grant is paid by the local education authority – what you get depends on where you are going to university or college, and how wealthy your parents are.

The maximum you can qualify for in 1991–2 is £2,845 if you are going to be living in London, £2,265 elsewhere and £1,795 if you will be living at home. However, that figure will be reduced if the parents' income is more than £10,600 after deductions such as mortgage payments and pension contributions have been taken into account.

If that residual income (as it is known) is more than £10,600, the grant will be reduced on a sliding scale up to £24,000. Once it reaches that level, no grant at all is paid.

These figures will not be increased in future. But students will be able to borrow money, instead. Interest, which is index-linked to inflation, is charged from the day you draw down the loan.

How much you can borrow depends on where you study, which year you are in and whether or not you live at home.

	Away in London	*Away elsewhere*	*At home*
Full year	£830	£715	£570
Final year	£605	£525	£415

Accommodation

Your biggest outgoing will be accommodation, if you are living away from home. First-year students are probably better off trying to get into the university or college halls of residence while they are adjusting to the new environment of student life. You will know in advance what you will have to pay; gas and electricity are normally included in the accommodation fee, so you won't be faced with a huge fuel bill during the winter; and other expenses such as insurance are lower for students in halls of residence.

If you decide on flat-sharing, you may need quite a bit of cash up front. One month's rent in advance is the norm, and you may also have to pay a deposit against breakages and damage to contents. (Tip: don't sign for any item the landlord promises to provide, it may never appear.) You may also have to pay a deposit for gas, electricity and the phone – and these can be up to £100 each – though you will get this money back when you leave.

Tip
If you are in dire straits, try going back to whoever gave you your grant and see if they have a hardship fund. Not everyone will be able to help, but it is worth a try.

Insurance

One thing you don't need when you are a student – if you are unmarried with no children – is life insurance. So don't let some smooth-talking salesman persuade you otherwise. If you want to save, use a building society or bank account. Life insurance has its uses, as life cover for people with dependants, but it should not be used as a saving scheme. Most of the first year's premiums go as commission to the salesman so if you cash the policy in early, during the first four years, you will get very little back.

Contents insurance for all your worldly goods is another matter entirely and a very good idea. No doubt you will be taking quite a lot of things such as a radio, stereo system, camera, computer, sports equipment, bicycle and so on with you when you go to college and it is unlikely that your parents' household insurance will cover you (though it is worth checking if you can have an add-on to that policy as it might be the cheapest way of covering your goods).

Students lose over £1 million of goods every year due mainly to burglars breaking into their digs or college room. So the £20 to £30 it will cost you to insure your possessions will be money well spent. Use a company which specializes in student insurance such as Endsleigh Insurance Services or some of the High Street banks, as they offer the best deals.

Students in official halls of residence can insure up to £1,750 of possessions on an 'indemnity' basis (that means the insurance company will take wear and tear into account when paying out on any claim) for around £20 a year. If you want 'new for old' (where the insurance company replaces your stolen or damaged goods with new ones) that would add an extra £5 to your annual bill.

At the other end of the scale indemnity cover for the same possessions if you live in a city centre, where burglaries are more common, would cost £50, new for old would set you back £55.

Tax

Like everyone else, all students have a personal tax-free allowance that they can make use of. But they only have one. So once income, from whatever sources (except any local or education authority grant, parental contributions, scholarship or research award), breaches the allowance, students again like everyone else have to pay income tax.

Vacations

Students working in their summer holidays won't have to pay tax if their earnings are lower than the personal allowance.

If that is the case, sign Form P38(S) (available from your local PAYE tax office), give it to your employer and you will get your earnings without tax having been deducted. If you have worked before, or claimed unemployment benefit, you will have a P45 – give that to your employer instead and he will be able to work out whether or not you will have to pay tax. If at the end of the tax year you think you have overpaid your tax, claim back the extra using Form R49(S).

Whatever you earn – providing it is more than £54 a week – you will have to pay national insurance, and that will be deducted automatically by your employer.

If you don't get a holiday job during the long vacation or it doesn't last the full time, you may be able to get some help from the Department of Social Security.

There are three types of benefit:

- *contributory benefits* (including pensions and Unemployment Benefit); you have to have paid enough NI contributions to receive these
- *non-contributory benefits* (including Child Benefit): you have to satisfy certain requirements to receive these, but they are not based on NI contributions
- *means-tested benefits* (including Income Support and Housing Benefit); you have to give details of your income and outgoings and you will get help if your income is low enough.

Students normally get help through Income Support for the long summer vacation if they do not have a job – though they will have to show that they do not have enough money coming in from other sources to live on. If you want to claim, use Form B1 from your local unemployment office. Students getting Income Support may also get Housing Benefit from their local authority to help with the rent.

If students have a good enough NI record of contributions, they may get Unemployment Benefit.

Students can't claim benefit for the short vacations as the grant is designed to cover those periods.

FIRST JOB

Getting your first job is exciting – getting your first pay cheque can be overwhelming. Particularly when you realizc that there are certain stoppages taken off your money before you even touch it.

You'll have to pay income tax, National Insurance contributions and perhaps even money into the firm's pension scheme (though that would be unusual for school-leavers).

The amount of tax you pay in a year is spread evenly over the twelve months and it is done by using a PAYE code (Pay As You Earn). To crack your tax code see Chapter 7, Understanding Income Tax, and to understand your pay slip see Chapter 8, Tax and Your Job. It is important that you do understand it, otherwise you won't know if you are being paid the correct amount every week or month.

When you start work, particularly if it is towards the end of the tax year on 5 April, you may not pay income tax at all if it looks as if you will earn less than the personal tax allowance for that year.

But you will pay National Insurance contributions – these are deducted by your employer automatically before you get your salary. How much you pay depends on how much you earn.

The classes of National Insurance contributions, and the amount you pay, are detailed in Chapter 7, Understanding Income Tax.

You will get a National Insurance number automatically before you leave school. It will have two letters, six numbers and a final letter, and will look like this: AB 12 34 56 C. You will probably get a plastic national insurance number card with the number on it. This is your national insurance number for life so you might as well try to memorize it. Every employer will want to know it and virtually every DSS form you fill in will need your NI number.

Bank accounts

Before starting your first job, it is a good idea to open a bank account. Most employers will offer you the chance to have your salary paid directly into the bank (indeed some employers will insist on that) so it helps if you have your account number and the address of your branch handy. If you open a current account, you will get a cheque book and generally a cheque card, but until you reach the age of eighteen you won't be able to overdraw or have a credit card because you are not legally responsible for your debts. With some banks, your cheque guarantee card is also your credit card. To stop you using it as a credit card your credit limit will be kept at £1 until you are eighteen.

> **Tip**
> No matter how little you are earning, try to save a pound or two regularly every week. It is a good habit to get into and it will mean that you are building up a savings record. When you come to want a loan it will be easier to obtain and it may be slightly cheaper, because you can show that you have managed to put a little bit of money together yourself. It is not always the size of the capital sum that the bank or building society manager will look at, but the regularity with which it was saved.

UNEMPLOYED SCHOOL LEAVERS

If you don't get a job when you leave school, there are various options open to you. Your school will outline most of them to you but, as a rule of thumb, you can either go on some sort of training scheme or go on the dole.

There is some talk of cutting the dole money of youngsters who won't take up the offer of job experience or training on a government-backed scheme.

Sixteen-year-olds

Sixteen-year-olds are currently eligible for the Youth Training Scheme. The scheme gives them two years of training – part of which will be off-the-job training in a college – and pays:

- £29.50 a week in the first year
- £35 a week in the second year.

If your travelling costs amount to more than £3 a week, they will be paid for you.

In Northern Ireland it is called the Youth Training Programme.

Youngsters should also make sure they keep a National Record of Achievement while they are doing this training. It is a *This is Your Life*-style book – a red hardback which documents anything of interest or value that they do that might help them to get a job. No doubt it would start with any academic or sporting qualifications they got at school and move on to include any other achievements, skills or experience they gather along the way.

Seventeen-year-olds

Seventeen-year-olds are eligible for the Youth Training Scheme – one year's training both in work and in college and they will be paid:

- £29.50 for the first three months
- £35 for the next nine months.

Eighteen and upwards

You are now eligible for Employment Training. Priority will be given to eighteen-to-twenty-four-year-olds who have been unemployed for six to twelve months, though anyone who has been unemployed for over six months, whatever their age, will be eligible.

The scheme will give you training in a type of career, worked out by you and your training agent, and the pay will be your current benefit money plus £10 a week.

All travel costs over £5 will be paid. Child-care costs of over £50 a week per child, if you are a single parent, will also be covered.

Jobstart allowance

If you have been out of work for more than twelve months, and take a job paying less than £90 a week gross, you will get an additional £20 a week for six months.

If you can't get a job or a place on any training scheme, you may be able to claim Income Support.

Starting your own business

Many youngsters who can't get a job opt to work for themselves. If you have a skill or an idea that looks like it could earn you a living you can get help with putting it into practice.

Finance

Training and Enterprise Councils (TECs) have been set up nationwide under guidelines laid down by the government. If they think that your plan is viable, then they can pay you between £20 and £90 a week for between twenty-six and sixty-six weeks as a sort of salary. This allows you to get your business up and running before you need to take money out to live on. How much you actually get from your local TEC depends on your needs and the needs of other people in your area.

Business help

Your local TEC or Jobcentre should be able to tell you about local courses that will help you to gain some business know-how. Your local college or polytechnic might run courses to allow you to train in computers or business skills, or you could attend a day-long enterprise awareness event.

How to claim

If you are under eighteen you will have to register for work at your local careers office. The careers office will give you a card to take to your local Unemployment Benefit office. If you are over eighteen, go straight to the Unemployment Benefit office. You will get a claim form, Form B1, which you send on to the DSS. You will probably have to go to your local DSS office for an interview and the DSS will work out how much money you are entitled to.

The money will be sent fortnightly to you by Girocheque.

You can't get any benefit unless you are available for work. However, if you are studying on a course that lasts for less than twenty-one hours a week it will not affect your benefit.

21 Holiday Money

How to take your money on holiday – the advantages and disadvantages of the options

Wherever you go on holiday, don't be left counting the cost of the Costas. A little advance planning over how to take your money can save you not only a bit of cash, but a tremendous amount of trouble. After all there is no point in relying on cashing traveller's cheques in the local bank if you are going abroad for Easter – the banks won't be open.

There are choices galore when it comes to taking your money abroad; which you choose depends on where you are going and for how long.

THE OPTIONS

Cash

Advantages. Taking sterling is the easiest way of all to carry your holiday money. It doesn't cost you any commission to buy it and you don't have to convert it back when you return home. Always have some sterling with you in case of an emergency on your return and as a back-up fund in case you run out of money.

Disadvantages. There are few shops or restaurants overseas that will accept £1 coins or £10 notes in payment. Your insurance policy will only cover you for up to £200 or so of cash that is lost or stolen but they would expect you to show reasonable care and keep money safe. More than that would not be covered.

Tip

Don't be tempted to use the black market and change your cash or traveller's cheques into the currency of the country using a shady character on a street corner. At best, you won't be able to change anything you have left over back into sterling. At worst, you could end up in prison.

Where to cash. Change sterling into foreign currency at banks, bureaux de change or hotel receptions – but check commission charges and exchange rates before you part with your money.

Foreign currency

Advantages. It is handy when you get there if you need to buy something in a hurry, get a taxi to your hotel or tip the porter.

Disadvantages. If it is lost or stolen you will have to report your loss to the police – losing valuable holiday time – and you won't get a refund until you return home, even if the insurance policy covers the loss.

Where to buy. Banks and some building societies will sell you foreign currency, though there will be a commission charge, usually between £1 and £15 depending on how much currency you want.

Traveller's cheques

Advantages. They are easy to carry, you know how much you have got and they are replaceable immediately if they are lost or stolen; that is why most package-holiday travellers take their spending money in traveller's cheques. (If the traveller's cheques company thinks you may not in reality have lost or had your cheques stolen, they won't refund the cheques to you while you are on holiday.)

Disadvantages. You have to buy traveller's cheques in advance so you have to know how much you will be spending on holiday and you will lose the interest on the money while you are away.

Where to buy. You can buy them at banks, building societies or travel agencies and they usually cost 1 per cent of the sum you are buying (that is £1 for every £100) for sterling traveller's cheques and 2 per cent if they are foreign currency cheques, with a minimum charge of £3.

Some banks and building societies sell traveller's cheques to savers free of commission so it can be worth opening an account to qualify, while some travel agents will sell you your traveller's cheques commission-free if you are booking your holiday through them. The saving will buy your first holiday drink.

Where to cash. Changing sterling traveller's cheques can be a bother. You will get the best rate of exchange in a bank – but you will be charged a commission. You won't be charged a commission for cashing the cheques in the hotel or in shops or restaurants but you will probably get a poor rate of exchange.

> **Tip**
> In some countries, such as Holland, Belgium, Austria, Malta, Italy, Denmark, Norway and Sweden, the banks charge commission on the number, rather than the value, of cheques cashed so take large denomination cheques.

You can buy traveller's cheques in sterling or foreign currency. If you are travelling to America always take your traveller's cheques in dollars because you will then be able to use them just like cash. You can spend them in restaurants, shops, etc., and you will get change from them just as if you had handed over a $10 or $20 bill. Don't do as I did on my first visit and try to cash a sterling traveller's cheque. Bartering with moon rocks may be possible, cashing a traveller's cheque made out in pounds is not – at least not without an enormous, and unnecessary, amount of difficulty.

In Europe you are often better taking your traveller's cheques in the currency of the country you are travelling to because you don't have to pay commission when you cash them – but if you bring any home with you, you'll have to pay commission when you cash them here.

Remember. When you buy your cheques you will be given a leaflet explaining what to do if the cheques are lost or stolen. Keep the leaflet and the cheque numbers somewhere safe, and you will indeed get your cheques replaced if you have a mishap.

Eurocheques

Advantages. The Uniform Eurocheque System is one of the easiest and most accessible of schemes. All the major High Street banks are now members so if you have a bank account you can apply to join.

If you apply – and do so at least three weeks before you travel – you will receive a plastic Eurocheque guarantee card and a wallet of cheques. You write the cheques in the currency of the country for any amount you want up to the equivalent of £100, sometimes more, and the card guarantees the cheque. If you want to spend more than £100 at one time, you just write more than one cheque, and they will all be guaranteed. The cheques are acceptable in Europe in hotels, garages, restaurants and shops, and the money comes out of your current account back home. But you don't have to pay immediately; the money is not taken out of your account until the cheque arrives back in this country.

Depending on which bank you are with, you might be able to use your Eurocheque card overseas in an ATM. Midland pioneered this, and it has proved so popular that many other banks are following suit.

Eurocheques are a super idea for anyone travelling around in Europe who doesn't know in advance how much they will be spending in each country.

Disadvantages. The drawback is the cost. The card costs £6–£8 depending on where you bank, the handling charge for each cheque is around 30p and there is maximum 1.6 per cent commission on the value of every cheque with a minimum charge of around 90p. They are a very expensive choice for a family having a two-week package holiday on a beach.

Where to buy. Apply through your bank branch.

Where to cash. Anywhere in Europe where you see the Eurocheque symbol.

Credit cards

Advantages. You don't have to pay the bills until you return home. Credit cards are acceptable in thousands of shops, restaurants, garages and hotels worldwide and are very useful if you run out of funds.

You can use your card to get cash in most European countries and to get money out of 'hole-in-the-wall' machines in some European countries.

Disadvantages. Not everywhere accepts credit cards. Most banks make an annual charge for their credit cards, and if you use your card to get cash there will be a charge. Access starts to charge you interest as soon as they know you have got the money but there is no commission charge.

Barclaycard/Visa charge no interest, but there is a commission charge of 1½ per cent. So when the monthly rate of interest is more than 1½ per cent use your Barclaycard if you have a choice.

If your credit cards are stolen they won't be replaced until you return home so don't rely solely on them.

Where to buy. Apply for a credit card using a special application form, or through your bank branch, at least a month before you travel.

Where to cash. Use credit cards instead of cash where you see the Visa (Barclaycard) or Mastercharge (Access) signs.

Tip
If you intend using your credit cards a lot, you might have to ask for your credit limit to be raised. Do this well in advance. Write to the credit card company asking them to raise the limit and explain the reason. If your credit rating is good and you tend to pay your bills on time, they will probably oblige.

Charge cards

Advantages. You defer paying the bills for a few weeks and charge cards can be replaced at short notice if they are stolen or lost.

Disadvantages. You have to pay for the cards and pay your charge card bill at the end of the month.

Shops and restaurants may try to charge you more if you use a charge card to pay the bill. They should not do that, but it can be difficult to argue your case.

Where to buy. Apply direct to American Express or Diners Club.

Where to cash. Where you see the American Express or Diners logo displayed.

The post office

Advantages. Girobank customers can buy Postcheques to withdraw cash in local currency from post offices all over Europe and North Africa. You'll need a cheque card, which is free but which will take up to two weeks to process and the cheques cost £6 for a book of ten. You can write as many as you like in a day up to around £100 each. Ideal because post offices open longer hours than banks and you often find a post office where you might not find a bank.

Disadvantages. You need a Girobank account.

Where to buy. Customers should apply to the Girobank two weeks before they travel.

Where to cash. Post offices.

No one is likely to use every option so choose the ones that suit you best. Most holiday-makers will want to take some foreign currency and perhaps a credit card if they have one. Whether you choose traveller's cheques, Postcheques or Eurocheques depends on where you are going, how much you are going to spend and whether you want to set yourself spending limits in advance.

Recommendations

- Package holiday in Europe – sterling or foreign currency traveller's cheques, foreign currency, credit card.
- Travelling holiday in Europe – Eurocheque card, credit cards, foreign currency, foreign currency traveller's cheques.
- Hitchhiking or camping holiday in Europe – Postcheques, foreign currency, credit card.
- Package holiday in America – credit and charge cards, US $ traveller's cheques.
- Travelling holiday in America – credit and charge cards, US $ traveller's cheques.

22 Your Personal Budget

Working out your personal budget · Coping with the big bills · What's on offer from banks and building societies · Banking from home

See also

- *Chapter 23, Borrowing Money*
- *Chapter 24, Coping with Debt*

Most people could do with more money every month – it is a bit like thinking you could cope better if there were twenty-five hours in every day. The solution to both problems is much the same: you can get by with what you have got *if* you just organize your money – and your time – better.

For your finances, what you need are a wet Sunday afternoon and a blank sheet of paper. It is time to organize your money.

This can be a really boring job if you are not the type that makes lists and likes to have your life organized. But it is worthwhile. Because if you don't understand where your money comes from and, more importantly, where it goes every month, you won't be able to take the next steps into the world of money.

But a few hours spent now, getting your finances out of the red and in the pink – if not entirely into the black – will pay dividends later on.

Why budget?

Mention the word 'budget' and you probably think of the Chancellor of the Exchequer holding up his tatty red dispatch case every year. Well, you are not far out. What you are about to do to your finances is a small-scale version of what he does for the country's finances every year. You're going to balance the books.

Find out how much you have coming in, and how much you have going out every year. That way you should be able to plan ahead and not be caught short by the big bills. Everyone knows that the fuel bills you get in the spring will be higher than the ones you get in the autumn because the winter is colder than the summer. So why do we never have a bit of extra cash budgeted for the spring gas and electricity bills?

I think the answer is to do a budget every year – call it the Day of Reckoning if you like. Circumstances change and you'll find that, by having a look every year, you will keep abreast of your cash flow.

A young single person spends more on holidays; young marrieds spend a higher proportion on their mortgage; once a baby comes along you have to budget for cots, and the loss, whether temporary or permanent, of the wife's wage. Then there could be school fees, educational trips abroad, teenager's clothes, maybe even a wedding to pay for. So, like the Chancellor, try to make your budget day an annual event.

THE BUDGET

The easy part about writing a personal budget is working out your income. Add together your salary and that of your spouse (if you are married), any interest you get from savings or investments, and any other income you have – maybe from shares left to you by an old aunt.

The difficult part is totting up the expenses. Try to fill in the spending side of the budget planner (see p. 249) as accurately as possible – and cross your fingers that it comes to less than the income side. You will need to add in one large item which I've called the replacement item because, every year, you have to replace something around the house. Whether it is the washing machine that has broken down, or the vacuum cleaner, or you need a new sofa or carpet, you will always have at least one large unexpected item to pay for. So you might as well budget for it.

The real joker in the pack is the 'extras' item. It has to include what you spend on presents, stamps, books, outings, prescriptions, babysitting, going to the theatre or cinema or zoo, money for school trips, vet's fees, taxis, cigarettes and so on.

Tip

If you find yourself overspending when you go out, here's a handy way to cut down. Carry less cash. Sounds simple and it works a treat. If you think you will need £20, take £15; if you think you'll need £7, take £5. You'll manage, and you'll make a saving too.

SPENDING

Fixed	**Everyday**
Mortgage	Supermarket/shops
Council charge	Clothes
House insurance	School meals
Gas/electricity/coal	Children's treats
Phone	Papers and milk
Season ticket/fares	Petrol/bus fares
Road tax	**Occasional**
Car insurance	Garden
TV licence	Holiday
Savings	Hobby/sport
Regular plan	Christmas and birthdays
Life insurance	Replacement item
Loan interest	**Extras**
Sub-total	**Sub-total**

Total spending

INCOME

Yearly salary	
Pension	**Total spending**
Child and DSS Benefits	
Interest and dividends	**Total income**
Anything else	
Total	**Surplus/deficit**

Personal budget planner

The problem is that it is the 'extras' which bring you down financially.

You can always remember the big expenses and work out in advance if you can afford them, but the trivial purchases you don't think about. And it is often these that clobber you at the end of every month. For example, you can probably remember how much you spent on the birthday present you most recently bought. But do you also recall paying another pound or so for the wrapping paper and card?

So treat this area a bit like a diet but instead of counting the calories, count the pennies.

If you find that money just trickles through your fingers, then it is a good discipline to see where it is going. Force yourself for a whole week to write down absolutely every penny you spend. You'll see at a glance what the rogues are – and you may be able to cut them down.

Perhaps you take more taxis than you should, or you spend a small fortune on endless cups of coffee at work, or your magazine bill is enormous. I'm not suggesting that you can or should cut them all back savagely, but at least be aware that these are the areas where you are spending more than you realized. And the time. If you use your lunch hours to troll the shops, you could cut down on what you spend by going swimming instead.

You'll also find that the very fact that you are forcing yourself to write everything down means that you will spend less that week. Your first saving. And if you go out and buy a new notebook and a sharp pencil to do this exercise, you are in a worse state than you thought!

Surplus or deficit

Dickens's Mr Micawber reckoned that if you had a pound coming in and you spent nineteen shillings and sixpence you were a happy man, but if you spent one pound and sixpence you were an unhappy man.

A surplus means your income is greater than your spending. It also means you should be able to cope with any unexpected bill, and start saving with no problems at all.

A deficit means your income is smaller than your spending. You have a problem. If you cannot step up your income, you will have to cut your spending. Have another look at your spending pattern and see where the cuts can come.

You probably won't be able to dock much off the fixed outgoings. They tend to be bills that need paying come what may. But you might be able to clip the everyday spending and perhaps cut savagely into the occasional. But do something now otherwise you'll be in debt before you know where you are.

When the bills come in

The second part of budgeting is being prepared for the bills coming in. There is no point in just dividing your spending column by twelve and expecting that figure to be your monthly outlay. You will find that several of the big bills come at the same time.

January and February are always cruel months. Not only do you have Christmas and the New Year sales to pay for, but the first of the winter fuel bills lands on your doormat too. If you also have your annual house insurance policy to renew and the car to tax, then you could find yourself with a huge overdraft which will lead to bank charges and interest payments.

So start to do something now.

Go through the spending columns of your budget planner again and underline the items which are not regular monthly or quarterly payments. For example, you probably pay the council charge and house insurance once a year, the papers and the milk when you remember, and buy bits and pieces for your hobby when you need them.

These underlined items are the ones you want to spread throughout the year. You can do it one of two ways:

- try to change from annual to monthly payments
- move the bills so that they don't bunch into one quarter of the year.

Changing from annual to monthly payments won't change the total amount you pay but it will save you from getting a huge unexpected bill. Most local authorities will allow you to do that with your council charge and many insurance companies welcome it too for the insurance premiums. Sometimes there is a minor administration charge – check that it is not more than a pound or two or you won't make any saving.

Fuel bills can also be paid monthly. The gas or electricity board will total your previous year's bills and divide by twelve, and that is what you will be charged every month. At the end of the year, the actual cost of the gas or electricity you have used will be compared with what you have paid. If there is a surplus the gas board will send you a cheque, if there is a shortfall you will get a bill instead. The electricity board does it a different way and works the surplus or deficit into the following year's monthly payments. But both fuel suppliers will use the actual amount of fuel you used to work out the level of monthly payments for the following year.

You can also have a budget account for your telephone.

Tip

If you want to pay monthly instead of quarterly, apply during the summer so that the monthly payments start in the autumn. That way you get reduced bills in the winter and then pay more than you normally would in the summer, so the fuel industry is subsidizing you. If you do it the other way round and start the scheme in May you are paying for gas or electricity before you use it, so you, in effect, subsidize them.

Things like season tickets can't be broken down that way. The reason they are cheaper than daily fares is because you are buying in advance and putting up a year's money in one lump sum. But you can shift the month in which you buy your season ticket. If you don't like having to buy your annual season in December just when you need all your money for Christmas, just get a monthly or quarterly ticket next time. Then you can buy your annual ticket in January or March. Of course it will cost you money to do this because the shorter tickets are more expensive, but it might stop you overdrawing your bank account. That way you could save more than you spend.

Alternatively, ask your employer if he will make you an interest-free or low-interest loan which you could repay monthly from your salary.

And try to pay the milk and papers regularly; £4 or £5 a week doesn't seem too bad – but £20 a month does.

If all else fails you can buy stamps to pay your gas and electricity bills and to save up for your TV licence. Try not to. It means you are paying in advance for goods and services. You would be better to discipline yourself to pay the money regularly into a bank or building society account so that at least it is earning interest for you. But buying stamps is one step better than saving in labelled jars on the mantelpiece and ten steps better than not buying stamps and getting into debt.

> **Tip**
> If you move house, or you are a separated wife who wants to transfer the electricity account into her own name, you may be asked for a deposit of over £100. Opt instead to pay your electricity bill monthly by standing order and you won't have to put down a deposit.

Records

There is one other thing you should have which is crucial to keeping a tidy financial mind – and that is a filing system of some sort. I'm not suggesting you get office furniture and filing cabinets but, unless you keep all your financial records in one place, you haven't a hope. If your Day of Reckoning requires a week of finding all the necessary facts and figures you will never do it.

Whether you keep one big file, a series of smaller ones or a spread sheet in your computer, here is what you should have on it:

- name and address of the banks and building societies where you have accounts
- the name and number of all your accounts
- the name and number of any shares you have
- the name and number of any NS certificates and premium bonds
- your salary details
- your pension details
- the details of any life insurance, term insurance or endowment plans
- details of any other financial plans you have
- premium number and details of any insurance policies you have
- details of your mortgage
- details of any HP agreements
- details of any loan agreements
- any financial letters that you receive during the year – such as notification of your building society and bank interest rates, your tax code, your mortgage payments and so on.

Money calendar

If you have a lot of savings accounts and shares or gilts it is also worthwhile keeping a note of when interest and dividends are due. Write down the twelve months and mark against each month the interest and dividends which are due. Tick them off as they come in, and if one is missing you will know right away.

Money calendar	
January	
February	BT dividend
March	
April	TSB dividend
May	
June	bank extra-interest account
July	NS certificates mature
August	
September	BT dividend; Abbey National interest
October	TSB dividend
November	
December	NS investment account; bank extra-interest account

You could take that one step further and mark up a yearly calendar with all the payments that are due, and all the ones you have to make. You then know at a glance which are going to be the lucrative months – and which are going to be the financially barren ones.

BANK ACCOUNTS

For most people, the basic budgeting tool is their bank account. Around 95 per cent of the working population have a current account. It is funded by their monthly salary or weekly pay and the money leaves the account automatically to pay the mortgage or rent, or it is withdrawn using a cheque or plastic card.

The problem comes when you withdraw more than has been paid into the account – that gives you an overdraft. An overdraft can be a very expensive item and may wreck any hopes you had of a proper monthly budget.

> **Tip**
> Never run up an overdraft without telling the bank first. The interest rates on unauthorized overdrafts are comparable with credit card interest rates and that can be up to 10 percentage points higher than the rate of interest charged on agreed overdrafts.

With most of the major High Street banks nowadays, current accounts are operated free of charge provided you remain in credit, and they all offer interest on credit balances. That means you can pay in money, write cheques, use your plastic card for withdrawals, have as many

Money wall chart		
Month	*Outgoings*	*Incomings*
January	phone sales gas	
February	electricity TV licence skiing	BT dividend
March	car tax car insurance	
April	gas phone council charge Easter hols. water rates tennis-club subscription	TSB dividend
May	electricity garden	
June	season ticket house insurance sales	bank extra-interest account
July	gas phone holiday	NS certificates mature
August	electricity school uniforms	
September	union dues	Abbey National interest BT dividend
October	gas phone rates water rates	TSB dividend
November	electricity	
December	Christmas	NS investment account bank extra-interest account

standing orders and direct debits as you like and transfer money in and out from other accounts free of charge.

You'll get interest on the deposit, but not as high a rate as you'll get for, say, a building society saving account, so don't let the balance go too high.

If you overdraw, all of these services will be charged for (in addition to paying interest on your debt). Most banks operate on a three-month accounting period. If you overdraw by only a few pounds in that three-month period you can be charged for every single transaction in and out of your current account during that period.

Most banks charge between:

35p and 45p for cheques
25p and 35p for automated withdrawals
25p and 40p for direct debits and standing orders.

On top of that, if your bank manager writes to you to complain that you have taken an unauthorized overdraft, that will set you back a further £10 to £20. And anything else that your bank normally does for nothing will be charged for if you dip into the red. All banks now send out regular lists of charges, so you have no excuse, really, for running up a big bill. A fictional example of bank charges is given below.

High Street Bank

Alison Mitchell
Account number: 123456

20 cheques at 40p	£8.00
6 ATM withdrawals at 35p	£2.10
2 direct debits at 30p	£0.60
2 standing orders at 30p	£0.60
Letter from manager	£20.00
Status inquiry	£5.00
Monthly fixed charge	£2.00
Total	£38.30
Interest on overdraft	£0.90
Total bank charges	£39.20

Tip
If you only dip into the red once, and then only for a small amount for a short time, ask for your bank charges to be reduced. Computers work out the charges but real people deal with your letter. Real people are more sympathetic to this sort of bad luck.

Standing orders and direct debits

If you pay money out of your account regularly, you will save yourself quite a lot of effort by having the cash transferred automatically by the bank. All you do is notify the bank of who and where you pay the money to, what day of the month you want it paid, and how much should be sent. The transaction will be done in one of two ways:

- standing order
- direct debit.

Standing order. If you are paying the same amount every month or quarter or year to a company or person – such as a mortgage or subscription – then do it through a standing order. Give the bank all the details, including the amount of money you want to pay, and they will transfer the cash on the given day. If the amount changes – because, for example, the mortgage rate goes down or your subscription rises – notify the bank and they will change the standing order.

Direct debit. A direct debit allows a company or person to take money out of your bank account and is usually used for accounts that you pay regularly but where the amount can vary – for example gas, electricity and phone bills, and insurance premiums. A lot of people don't like this because they don't know how much is going to be taken out, although you must be given fourteen days' notice of a change. If you think a direct debit will get you into financial trouble, don't have one. The company can just bill you in the normal way and you send them a cheque.

Cheques

You can write a cheque for any amount – providing you have money in your account to cover it, or you have an overdraft facility. But if you want the cheque guaranteed – that is the bank will agree to pay the cheque regardless of how much you have in your account – then you will be limited to a maximum of £50 or whatever figure is written on the cheque guarantee card.

Your cheque guarantee card, which you will need to back the cheques you write in shops, restaurants, garages and so on, only covers you up to £50. If you are spending more than that you will need to have a very honest face, pay for it another way, or not expect to get your goods until the cheque has been cleared.

Most supermarkets now will take cheques for up to £100 provided you write your name and address on the back and can provide identification – other places might not. And don't think you can get round the rules by writing two cheques for less than £50 each. The guarantee won't stand and if you don't have the funds or the overdraft facility, they will bounce the cheques.

Some banks and building societies now offer higher levels on the guarantee cards of high earners.

Bouncing. If you don't have enough money in your account to cover your cheque, and your bank doesn't want to give you an overdraft or increase the one you already have, it will not pay your cheque. The cheque will be returned to the bank of the person who presented it – that is the shop or whatever that you wrote the cheque to – marked 'refer to drawer, please re-present'. At the same time the bank will write to you telling you the cheque has been returned because of 'lack of funds'. This gives you a chance to put more money in your account before the cheque is re-presented.

The bank will do this once or twice and the third time it will normally just say 'refer to drawer'. Then it is up to the person who presented the cheque to get in touch with you for a replacement.

It is known as 'bouncing' your cheque. And it will cost you. Not only will you be embarrassed by the cheque bouncing but there is a bank charge of around £7 to £20.

A cheque for less than £50, covered by a cheque guarantee card, cannot be bounced regardless of whether you have enough money in your account to cover it.

Stopping. If you lose a cheque or the person you give it to loses it, or the cheque is stolen, you can have it stopped. You should ring your bank with the details of the person the cheque was made payable to, the number of the cheque, the amount it was made out for, the date and your account number. The bank will not cash it when it is presented. You will then have to put all these details in writing. The charge for stopping a cheque is around £5 to £10.

You cannot stop a cheque for less than £50 that has been backed by your cheque guarantee card and has the number on the back. And if you stop a cheque for no good reason, then you can be sued by the person or company to whom you wrote the cheque.

Tip

If you regularly want to pay for goods in this country or abroad that cost more than £50, apply for a Eurocheque card. That guarantees special cheques (which still run off your current account) up to £100 each and you can write as many guaranteed cheques as you like for the same item. The drawback is that a Eurocheque and card does cost a lot more than an ordinary cheque. (See p. 243.)

Budget accounts

Most banks run some sort of budget account to help you with your financial planning. The accounts all differ in detail, but work on the same basic principle.

You add up all your major household bills for the year – or at least the ones you want to pay for through your budget account – divide the total by twelve and you pay that amount into your account every month. The account can then be used for all the bills included in the running total. Because not all bills are monthly you are usually allowed a credit limit of up to three or four months' worth of payments – for example, if you pay in £100 a month your limit could be £300 or £400. This covers the big quarterly bills you have to pay.

Of course you have to pay for an account like this and there is usually a flat fee agreed in advance and you may have to pay bank charges on the transactions. When you go into the red you will be charged interest on the money you owe the bank, usually at the normal overdraft rate.

How to avoid bank charges

- Stay in credit.
- Get bank statements regularly and check them.
- Don't pay bills earlier than you have to.
- If you are going to overdraw, write as few cheques as possible – use your credit card to pay the bills, and one cheque to pay off the credit card.
- Get a cash advance on your credit card to avoid overdrawing (though you will have to pay interest on this money of course).
- Keep a diary note of when standing orders and direct debits will hit your account.

Joint accounts

A joint account to pay the household bills is the answer for many people. Enough money is paid in each month, by one or both partners, to cover the general bills.

The accounts run in one of two ways. Withdrawals need:

- either signature
- both signatures.

It is up to you which you go for – the former is more trusting than the latter. But if there is a bitter marriage split there could be problems. One partner could withdraw all the funds, run up a huge overdraft and run off leaving the other partner liable for the debt.

Statements

Try and get bank statements on a monthly basis. They are often only sent quarterly unless you ask for them to be more frequent. But there is no point in getting a statement every month unless you check all the entries.

You will probably find that you go for months without spotting any error, but it only takes one rogue debit or credit to knock your budgeting for six. And if the bank pays money into your account in error you should tell them about it. Unless you can prove that you genuinely knew nothing about it – and that would be difficult if it is for a big sum – then you have to pay the cash back.

The other good thing about monthly statements is that it makes you spend half an hour or so every month keeping tabs on your finances. If you do that you'll know very quickly if you are going off the rails.

'Hole-in-the-wall' machines

Most banks offer plastic cards to go with their current and deposit accounts. These cards all have a special number to go with them, and the number is unique to each card. If you use your card and someone else's number more than twice, the machine will keep your card.

Armed with card and number you can then use the 'hole-in-the-wall' machines to do your banking business. With some banks you can pay in cash or cheques, withdraw from any of your accounts, transfer money or even leave messages for the bank manager.

But, watch out. If you lose your card or get it stolen and the thief also knows your number – some people are foolish enough to write the number on the back of their card – you could be liable for any cash that is withdrawn from your account.

Banks usually only enforce liability if the circumstances are dodgy, and they suspect you of collusion with the thief. Once the card has been reported as lost or stolen, there is no further liability to you.

If you don't think you will be able to remember the number you are issued with, try to change it. Some machines have a facility to allow you to tap in a new number for yourself. You probably won't be able to get 1234, or 0000 but there are plenty of alternatives you might find easy to remember. Your date of birth, or four of the digits from your telephone number for example. I have a friend who changed hers to 5050 because she thought that gave her a fifty/fifty chance of getting it right!

If you don't want a plastic card, or worry about it being used by someone else, cut the card up when it is sent to you, and return it to your bank with a letter telling them that you don't want it. Keep a copy of the letter so that they will never be able to charge you for cash that is taken out of the machine by someone else and charged to you.

Phantom withdrawals. Many people complain that money is withdrawn from their accounts by card and they neither withdrew the money nor did they lend or give their card to anyone else. Under the new Code of Practice, the onus is now on the bank to prove that you did use the card, rather than on you to prove that you didn't.

If the branch won't believe you, complain forcefully to head office and then to the Banking Ombudsman.

BUILDING SOCIETY ACCOUNTS

There was a time when you went to your building society for a mortgage and your bank for a cheque book. No more. The banks are strong in the field of home loans and some building societies can offer you a current account. The current accounts offer interest of a few per cent less than ordinary saving rates but you get a cheque book and guarantee card, and can have standing orders and direct debits operating from the account and an overdraft.

Large building societies offer PIN cards to go with their current and deposit accounts.

However, if you need some financial sophistication in your account, you may find the building societies lacking. For example, it can take up to ten days to clear a cheque.

HOME BANKING

There are two types of home banking:

- phone-based
- screen-based.

With the phone-based service you phone into a central computer to do your banking business. You can check account balances, transfer funds, pay bills, amend standing orders, order foreign currency and order bank statements.

The screen-based accounts also need a phone link, but they have the advantage of allowing you to see, and print out, your accounts and transactions.

Both types of accounts cost extra to run – the screen-based ones tend to be more expensive than the phone-based ones. Ask your bank for details of costs.

COMPLAINING

If you are dissatisfied with the service you are getting from your bank or building society, complain.

Start with your local branch and if that gets you nowhere write to the bank's head office. Your local branch will give you the address and usually the name of the special department which deals with complaints.

If your complaint is still not dealt with to your satisfaction you can use the ombudsman.

Banking Ombudsman

He will only deal with your case if you have tried, unsatisfactorily, to clear up the matter with the bank. He can order the bank to pay compensation of up to £100,000 and his decision is binding on the bank though not on you.

The Banking Ombudsman, Citadel House, 5–11 Fetter Lane, London EC4A 1BR (tel. 071 583 1395).

Building Society Ombudsman

He can order compensation of up to £100,000 and his decision is binding on the building society unless they state publicly their reasons for not accepting his decision, but not binding on you.

The Building Society Ombudsman, Grosvenor Gardens House, 35–7 Grosvenor Gardens, London SW1X 7AW (tel. 071 931 0044).

23 Borrowing Money

Where best to borrow money – all the options · Comparing interest rates · Credit scoring · Credit reference agencies

See also

- *Chapter 24, Coping with Debt*

Not everybody borrows money because they can't afford to pay cash. Some people prefer to take out a loan rather than use up all of their savings if they are buying something big; others use credit cards and charge cards when they are shopping but pay off the bill at the end of every month. But a high proportion of those people who do borrow money do so because they haven't the ready cash to pay for what they want.

Borrowing money to buy something can make good financial sense. If you see something you need at a reduced price you may save more on the purchase price than you will pay in interest charges. Or a large item may be offered on interest-free credit.

Borrowing money means you are taking on a debt and if you think you won't be able to handle the debt, don't borrow the money. There is nothing wrong with saving up first and then buying whatever it is you need or want.

It is crucial when taking out a loan that you get it right. Make sure you borrow:

- from the right place
- at the right rate of interest
- and over the right length of time.

Time

How long you take out your loan for depends really on what you are using the money to buy. If you get a loan for £1,000 to pay for a holiday you should really agree to pay back the money within twelve months. Indeed you need to think carefully about borrowing money over a longer period, otherwise you will still be paying back for the first holiday, after you've had (or need) the second.

At the other end of the scale, a loan for a house will probably be over a twenty-five-year period. No one would expect you to pay that one back within a couple of years.

But for everything else there is really no such optimum time. It is a good rule of thumb to try and make sure the loan will be paid off before the purchase has packed up. So, for example, if you are taking out a loan for a car try to make sure the life of the loan is shorter than the likely life of the car. Or you will still be paying for the car when you are back to taking the bus.

The longer the loan the cheaper the monthly payments – but the longer you have to pay them. And in the end the more you pay in total. So don't be taken in by the adverts which offer to take on all your debts and turn them into one, much lower, monthly payment. It is called consolidating your loans. It means you pay less each month, but for much longer, so you end up paying back much more. And if you had trouble with your debts in the first place this will only compound the problem, because you are turning a short-term debt into a long-term one.

Repayments on a £2,000 loan

	Repayment period		
	3 years	*7½ years*	*10 years*
Weekly payments	£16.23	£8.87	£7.77
Total paid	£2,532	£3,460	£4,040

Interest rates

All financial advisers will tell you that shopping around for a loan pays off – and it does. The difficulty is making the comparisons. After all if you see the same iron in two shops at two different prices, you will buy the cheaper one. But the price of a loan is more difficult to work out. If you wanted to borrow £200 over two years, which of these options would you choose:

- a flat rate 11 per cent bank loan

- 1.75 per cent a month
- £3 a week?

The answer is the bank loan. But on the evidence above it would be almost impossible to work that out. Luckily there is a magic formula. It is called the APR – the annual percentage rate. The APR is the true rate of interest and usually includes any charges, such as an arrangement fee, that are included in the deal.

How an APR is arrived at shouldn't really worry you – all you need to know is that the APR is the comparable rate. No matter how the rates of interest are quoted, compare the APRs because that is the price of the loan. All interest rates have to have an APR figure quoted next to them and, under the rules of the Consumer Credit Act, it should be in the same size of print – only a 'cash' price is allowed to be larger. If there isn't an APR quoted, don't take the credit or the loan. To take the above example again:

- a flat rate 11 per cent bank loan has an APR of 21.7 per cent
- 1.75 per cent a month has an APR of 23.1 per cent
- £3 a week has an APR of 61.2 per cent.

So with the APR written in, it is easy to see which is the best deal.

As a rule of thumb, loans that you set up in advance are a better deal. The lender has a chance to check you out financially and he will only take you on if he thinks you are a good risk. That means that it is a good bet you will pay back the loan. And if the lender is only going for good risks, the interest rates will be lower.

If you get credit over the counter, the checks made on you are much less thorough so there is a percentage built in as a cover for bad debt. If you are a good risk, then you lose out by paying higher rates of interest unnecessarily.

And if you offer security, or collateral as it is officially called, you will probably get still keener rates of interest. That is why it can be worthwhile offering your house as security for a big loan. But remember if you do that, and fail to pay off the loan, the person who lent you the money can sell your house to get the cash. Tread warily when you tread that path.

TYPES OF LOAN

Overdraft

What it is. An overdraft is a cheap way to borrow money short term. It is not a formal loan, you just spend more than you have got in your current account. It is used most to tide people over from one month's salary to the next.

Interest. You are charged interest on a daily basis. Interest rates tend to be lower than credit card rates.

Drawback. If you overdraw on your current account, for however short a time, you can make yourself liable for up to three months' worth of bank charges. And if you overdraw without the prior consent of your bank, you will pay an extra ten percentage points in interest charges. Most banks charge a fee for arranging an overdraft.

Availability. Overdrafts are available on your bank current account, and some building society cheque accounts.

Personal loan

What it is. A personal loan is a good way of borrowing money over a set period of time – usually one year, two years or five years.

Interest. The rate of interest will be quoted to you as a flat rate. That means if the flat rate quoted is say 10 per cent, the interest will be £10 per £100 borrowed per year. As a rule of thumb the APR is usually a little less than double the flat rate quoted. So a flat rate of 10 per cent would have an APR of around 19 per cent. If you are borrowing in the short term, say under six months, an overdraft will be cheaper, but over the longer term you will do better with a personal loan.

Drawback. Personal loans have to be set up in advance and as such are no good for spur-of-the-moment buys.

Availability. Banks and large building societies offer personal loans to customers and non-customers – though you would have to have some financial credentials such as a house or job.

Revolving credit account

What it is. It is an account that you can use to save in and borrow from. How much you can borrow depends on how much you can pay in each month – the borrowing limit is usually up to thirty times what you agree to pay monthly. It is more flexible than a personal loan because you don't have to know in advance exactly how much you want to borrow, nor do you need to tell the bank what you want the money for, and as you top up the account every month you can borrow again. The account has its own cheque book.

Interest. The interest on the money you borrow will probably be the same as the personal loan rate. The interest on any credit balances will be the same as the lowly bank deposit rate.

Drawback. The interest rate on any money you have in credit is low, so if you find you are saving rather than borrowing use a proper savings account. This type of account can encourage you to spend up to your limit.

Availability. Most banks.

Credit cards

What it is. Credit cards are plastic cards – Access and Visa are the most common – which you can use to pay for goods and services. You only have to pay back £5 or 5 per cent (whichever is the higher) of the total every month. If you do pay the full amount, you won't be charged any interest so you can get up to six weeks' free credit. That means from the date of buying the goods to the date of paying the money could be as long as six weeks. In practice most people get around three weeks' free credit.

Every customer has a pre-set spending limit and once that is used up you can't charge any more to the card until you pay off some of it. You can use credit cards in this country and abroad.

Credit cards are a good way to spread your shopping bills and are good value interest-wise if you are only borrowing for a few months.

Interest. Most credit cards tend to charge roughly the same rate of interest. The interest rate is quoted on a monthly basis, for example 1.8 per cent a month, which would be 21.6 per cent a year. This is an APR of 23.8 per cent. Each statement has two dates:

- the date the statement is sent out
- the date the money is due into the credit card company.

If you pay in full every month, you don't pay any interest. If you don't pay in full you are charged interest. And nowadays most credit cards make an annual charge of £10 to £15.

Drawback. If you are not careful, credit cards can lull you into a feeling of false security and encourage you to spend more than you would if you had to pay in cash.

Availability. Most banks can issue you with a Visa and/or Access card.

Charge cards

What it is. A charge card is very like a credit card, but there are no pre-set spending limits, and you have to pay off the full balance every month. The two main groups are American Express and Diners Club but most banks also run their own up-market charge cards too. If you have a gold card or platinum card you can get an overdraft facility from your bank to run alongside your charge card. You use the overdraft to pay the charge card bill, and pay off the overdraft at your convenience.

Interest. There is no credit on an ordinary charge card so there is no rate of interest. If you have a gold card your overdraft rate will be at a slightly lower rate than you would have to pay for an ordinary overdraft.

Drawback. Not everyone qualifies. You have to be a high earner to apply.

The cards cost money – both to join and for the annual subscription. If all you get out of having the card is a status symbol, stick with the credit cards.

Availability. Apply to American Express, Diners Club or your bank.

Debit cards

What it is. A debit card is really an alternative to a cheque book and card. They don't offer credit, all they do is give you an alternative way to pay. And, of course, there is no £50 limit on your spending.

Interest. If you have money in your current account, there is no interest to pay; the cash is just deducted from your balance. Otherwise it will be overdraft rates.

Drawback. They are not as widely accepted as cheques because they need a special machine to operate them.

Advantages. There is no £50 limit to your spending, and the cards can be multi-purpose, operating as cheque-guarantee cards as well and operating cash dispensers.

Store cards

What it is. Store cards are credit cards run by large department stores or groups. You buy your purchases and charge them to your account and at the end of the month you get a bill, which you can pay off the same way as a credit card. Marks and Spencer, which runs the largest of the in-store credit cards, call theirs a charge card because customers charge goods on to it. But it isn't, it is a credit card because you don't have to settle the bill in full at the end of the month.

Interest. Interest rates will be quoted monthly. Some are reasonable and tend to be similar to what other credit cards charge. Others can be almost double, so watch out.

Drawback. Because you can open a store card account over the counter and get quite a high credit limit, they can tempt you to spend money you haven't got. If you shop in a lot of places you will need a thick wallet for all the cards. And, of course, you are limited to shopping in one store or chain. That could make getting the best buy difficult.

Availability. Most large High Street stores.

Hire purchase

What it is. Hire purchase is the most basic form of impulse buying. You see something you want and cannot afford, and you are offered instant credit so that you can take it away with you. Very useful if you don't have the time or energy to set up a bank loan in advance, or you don't have the financial credentials to qualify for one.

Interest. Interest rates tend to be quoted as monthly payments – for example £7.22 a month – but there will always be an APR quoted next to it and these can be very high indeed. If you want something quickly and you think you can afford the monthly or weekly payments, you might ignore the APR – but that is a mistake.

Drawback. Because HP is offered to almost everyone who asks for it, the rate of interest has to be high to cover the proportionately higher number of bad debts that the HP company takes on. Don't take out HP without first checking that the APR is at least within shouting distance of a bank loan APR.

Availability. Most High Street shops, particularly the large electrical retailers.

Mail order

What it is. If you buy something from a picture in a catalogue, magazine or newspaper and the goods are sent to you, that is mail order shopping. The mail order catalogues have updated their image dramatically over the past few years and their prices are now much more competitive. You don't always pay for your goods when you order, but when you receive them and decide to keep them. You can pay in one lump sum, or spread the cost over a certain number of months.

Interest. No interest is charged.

Drawbacks. Financially, there are no drawbacks to taking interest-free credit if the goods cost no more from the catalogue than they would from a nearby shop.

Availability. From catalogues.

Credit unions

What it is. A credit union is a super alternative to moneylenders, pawnbrokers, HP and the like. They are a little like non-profit-making banks. If you have savings you pay them in; if you want to borrow you can get a loan from the credit union at very low interest. Savers buy £1 shares (rather than put in deposits) and get a dividend (rather than a rate of interest) and that can vary from 1 per cent up to an 8 per cent maximum. You usually have to save for a bit before you will be allowed a loan and there is a maximum that you can get – £2,000 more than your savings – but each credit union will have its own rules.

Credit unions are a growing movement and are set up and run by groups of neighbours, workers or members of the same church. Usually the members are people who might have difficulty getting credit elsewhere. All credit unions are covered by a bond which would refund the members' money should the treasurer run off with the funds! So they are quite safe.

Interest. Interest on loans is usually 1 per cent per month on a reducing balance – and there are very few bad debts.

Drawback. There may not be a credit union near you. If not, start one up in the area where you live or work.

Availability. Limited at the moment, but further information from the Association of British Credit Unions, Unit 307, Westminster Business Square, London SE11 5QI, and the smaller National Federation of Credit Unions at Fifth Floor, Provincial House, Bradford BD1 1ND.

Pawnbroker

What it is. A pawnbroker's is a shop where you take something you own to act as surety or collateral for the loan you will get. It can be a very small item such as a watch or radio, worth £5 to £10, or something much more valuable than that. And you borrow the money for up to six months.

Interest. Because of the flat-rate fee or ticket charge that most pawnbrokers operate, the APR can be as high as 40–50 per cent. On smaller items, where the ticket charge is worth a large percentage of the value of the goods, the APR can be even higher than that.

Drawback. If you don't redeem your goods at the end of the six months, the pawnbroker will sell them and if the money advanced to you was less than £25, you won't get anything from the sale. Interest rates are crippling and the money you get from the loan won't be anywhere near the value of the goods you have deposited.

Availability. Most large towns and cities, where you see the sign of the three gold balls.

Moneylenders

What it is. A moneylender tends to lend his own money – usually to people who have been refused credit elsewhere.

Interest. Very high indeed because of the risk taken by the lender.

Drawback. It is a difficult loan to default on and the interest rate will cripple you.

Availability. Moneylenders usually advertise in local papers and shop-window noticeboards.

Warning. They are often associated with petty crime and may try using strong-arm tactics to get their money back. Avoid them if you can.

Loans from employer

What it is. A lot of companies and small businesses will help employees by offering a loan where it is needed. Often the money is for job-related expenses such as an annual season ticket. If you earn under £8,500 a year the 'perk' will be tax free. (See Chapter 8, Tax and Your Job.)

Interest. This will depend on your employer, but is likely to be at a low rate, or even interest free.

Drawback. If you leave the company you might have to pay the loan back in one lump.

Availability. It depends on your employer. If there is no formal scheme, it might still be worthwhile asking for a loan.

LOAN INSURANCE

It is possible to take out an insurance policy to cover your loan should you be made redundant, fall seriously ill or die. Most insurance packages will make the monthly payments for you for an agreed length of time while you are unemployed or off work sick, or pay off the loan completely if you die. The larger the loan and the longer the term, the higher the cost of the insurance.

Tip
Remember to make a claim. If you do fall ill or are made redundant go through all the loans you have taken out recently to check for this insurance and if you've got it, use it.

CREDIT SCORING

So now you know what the options are if you want to borrow money, and you can choose the one that suits you. Unfortunately it is not always as easy as that – the lender may not choose to make you a loan. Most lenders – whether it is a bank, a credit card company or an HP company – use the same method for working out whether or not you are a good risk. It is known as credit scoring.

What they do is allocate you points for all your financial advantages – and if you reach a certain level you will get the loan.

There are five main areas on which you are graded:

- job
- salary
- marital status
- home
- bank record.

Job. If you have a job, and your salary is paid directly into the bank, you will get more points than if you are paid weekly in cash. You will also fare better if you have been in your present job for more than a year and if you are a supervisor or manager.

Salary. The more you earn, the more points you get.

Marital status. A married person gets more than a divorced person, who gets more than an unmarried person.

Home. Home owners are the tops. The longer you have lived at the same address the more points you get. There are certain housing areas, mainly in large cities, which are virtually outlawed by credit companies as being very high risk. Their experience of lending money to residents has been very bad indeed. It used to be known as 'red lining', which meant that the credit companies put a red line round the area and wouldn't lend to anyone within it. That practice no longer exists but there are still certain areas where they will take a closer look at you if your address falls within it.

Bank record. If you have a bank account you get more points than if you don't. And if you've had it for a few years and had no problems running it that is better still.

You will get minus points if you are unemployed, move address a lot or have a County Court judgement against you for whatever reason.

Of course credit scoring can throw up some odd anomalies. I would be refused a card by some companies because I don't own a house (it is in my husband's name) and I don't have an employer to verify my salary (I work for myself).

But I'm in good company. The Duke of Edinburgh, with no house of his own or salaried employment, would be hard pushed to qualify too. So if you are turned down because you don't suit the credit scoring system – but you think you are still a good risk – write and tell them the real score.

Credit reference agencies

Most traders don't operate their own credit scoring – they ask credit reference agencies for a report on anyone applying for a loan. Credit reference agencies collect information about people's financial standing and will send their report to the trader who pays them for this information.

If you are turned down when you try to hire something, like a TV or video, or for credit, and you think it is because of the credit reference agency report, then you have a statutory right to see that report.

Ask the trader which credit reference agency he used, and for the address. You must ask within twenty-eight days of being turned down. Write to the agency, asking for a copy of your report and enclose a fee of £1. The agency will send you a copy of your file, and a statement of your rights, which describes how you correct any mistakes on the file.

At this stage it is worth getting a copy of the Office of Fair Trading leaflet entitled 'No Credit? Your Right to Know What Credit Reference Agencies are Saying About You', which explains exactly what you do to get wrong information taken off your file. You can get the leaflet at your local CAB or library or directly from the OFT, Field House, 15–25 Breams Buildings, London EC4A 1PR (tel. 071 242 2858).

If the information on your file is not correct, then do something about it. If the credit reference agency won't change the details you can insist that they add a page from you explaining the mitigating circumstances.

Watch out

There can be lots of pitfalls around when you borrow money. Watch you don't get caught.

- Don't take lightly a request to act as guarantor on someone else's loan. If they default, you have to pay up.
- If you get credit from a credit card or HP company and the item you bought costs more than £100, you have a claim against the finance company if the goods or services were faulty. It is known as joint liability and it means that they have to cough up if no one else will. So if the trader won't help, or goes broke in the meantime, you might have a claim against the credit company.
- If you realize after borrowing money that you are paying extortionate rates of interest you can take the matter to court. Ask your local CAB or Trading Standards Officer for advice first.
- If a debt collector puts undue pressure on you – perhaps he is threatening to break your leg or is ringing you ten or fifteen times a day to ask for his money – he is breaking the law. Tell your local Trading Standards Officer or the police.
- Don't accept a quote for an interest rate without an APR to accompany it. If the credit company won't provide it for you take your business elsewhere.
- Unsolicited doorstep selling of credit is illegal, and probably expensive, so don't have anything to do with it.
- Always keep tabs on how much you have paid off, and how much you still owe. When companies get it wrong, it is rarely in the borrower's favour.
- If the trader is offering credit, you may have to pay the full price for the goods. Buyers with cash can often negotiate a discount. So organize a bank loan instead – to the trader you are now a customer with cash.

Your financial rights

You have a right to cancel a credit agreement if:

- you deal face-to-face with the trader

and

- you sign the form at home or anywhere else that is not the trader's premises.

A phone call is not a face-to-face interview.

If you and your husband both sign the form asking for credit, it is the last signature that counts. So if you discuss the deal with the trader and take the form home and sign it, and then your husband signs, you won't be able to cancel, because your husband didn't chat to the trader about it.

If you sign a credit agreement at home, and you have a right to cancel, this is what should happen: a few days after you have signed, a copy of the agreement will be sent to you and you have five days from receiving this second copy to cancel.

24 Coping with Debt

The signs of overspending · How to deal with debts small and large · Going to court

See also

- *Chapter 23, Borrowing Money*

Standing in the path of a charging elephant is not unlike getting into debt. If you're lucky you'll come out of the experience badly bruised and battered; if you're unlucky it'll overwhelm you completely.

No one sets out deliberately to provoke a full-grown angry elephant, and similarly, very few people set out to get into debt. It just creeps up on them.

Usually there are three paths that lead you into the jungle of the serious debtor:

- overspending
- an unexpected trauma
- a sudden jump in your expenses.

Overspending

If you spend more than you earn in one month, you will probably end up with an overdraft. If you continue to overspend, month by month, and take out more lines of credit than you can cope with to pay for all your purchases, then you will end up seriously in debt.

Unexpected trauma

Most people have debts of some sort – a mortgage, a personal loan for the car or the new kitchen, an overdraft and perhaps a couple of in-store credit card debts.

The problem for them comes when they are hit by an unexpected crisis. A sudden serious illness, marriage problems, redundancy, an unexpected pregnancy that means the wife's income is lost for at least a short time; all these might turn loans you can cope with into serious debt.

Rise in expenses

If you are sensibly balancing your family budget and keeping your outgoings just below your income, you won't have much left over to save. So a sudden and unexpected big bill will hit you hard. Moving house, a burglary you are not insured for, replacing the washing machine or even a funeral to pay for will lead you down the path towards serious financial problems.

A loss of earnings can lead you down the same path. Losing regular overtime or being put on to short-time working are more than enough to upset previously balanced budgets.

SIGNS OF OVERBORROWING

Overspending is like overdrinking. You don't realize the problem is serious until it is too late. So act now. Check yourself out to see if you are already on the slippery slope.

1 You see a new winter coat that's bang up-to-date in fashion. You can't pay by credit card because you have already reached your limit. Do you:
   (*a*) pay by cheque and hope your overdraft will cover it
   (*b*) open an in-store account to buy it
   (*c*) make do with last year's coat?
2 You are overdrawn at the bank and your bank manager is writing to you for the second time to complain. Do you:
   (*a*) put the envelope behind the clock unopened
   (*b*) go and see him to try and sort out the mess
   (*c*) buy yourself a new outfit to get over your depression?
3 You have spent up to the limit of your credit card and you are offered an increase in your limit. Do you:
   (*a*) turn it down so that you can't spend any more
   (*b*) accept with glee because it means you have money to spend again
   (*c*) turn it down – because you have applied for another credit card?
4 You are invited to a slap-up meal by someone you've been eyeing for

months. You are having a bad financial month but you'll need a new outfit. Do you:
   (*a*) skip paying the mortgage this month so that you can buy your outfit
   (*b*) wear an old favourite, but tart it up with new accessories
   (*c*) ask if you can go to the cinema instead?
5 You need a new washing machine and decide to buy it on HP. Do you pay the money back over:
   (*a*) 12 months
   (*b*) 24 months
   (*c*) 48 months?
6 You realize you are building up debts you are going to have difficulty repaying. Do you:
   (*a*) stay in for a month to save money
   (*b*) move all your debts into one large loan spread over a longer period
   (*c*) cut up all your credit cards so that you can't spend any more?

Answers

1.	(*a*) 2	(*b*) 3	(*c*) 1	4.	(*a*) 3	(*b*) 2	(*c*) 1
2.	(*a*) 2	(*b*) 1	(*c*) 3	5.	(*a*) 1	(*b*) 2	(*c*) 3
3.	(*a*) 1	(*b*) 3	(*c*) 3	6.	(*a*) 1	(*b*) 3	(*c*) 2

Score 6–8: you are unlikely to be bothered by debt problems. When the going gets tough you take all the right actions.

Score 9–13: watch out. You are on the verge of having money problems. If things keep going as they are you should be OK – but if you're hit by a crisis or an unexpected big bill you could go under.

Score 14–18: stop spending now. The signs of serious debt problems are all there so draw back. Have some lean months to get your finances back under control or you could end up at the bottom of the slippery slope.

Signposts to watch out for

- lots of small debts in different places
- several credit cards and in-store cards
- you have spent up to the limit on your credit cards
- you only repay the minimum on your credit cards every month

- your bank manager has refused to increase your overdraft any further
- your save and borrow accounts are always at their spending limit
- you don't open bills when they come in
- you have skipped paying the mortgage or rent on occasions
- you opt for longer loans so that you pay less back every month
- you have missed occasional HP payments
- you are beginning to think about taking out one large loan to cover all the small ones.

If you can say 'that's me' to more than three of these, you have a money problem. You must do something about it now because the sooner you act the more likely you are to survive.

COPING WITH THE DEBTS

Getting into debt is a lot easier than getting out of it again. The reason you overborrowed is because you didn't have enough money to pay your monthly bills. Now you not only still have regular bills, you have the interest on your loans and the capital to repay.

There are two types of debt:

- small
- large.

And you deal with them in two different ways.

Small debts

The small debt is much easier to deal with. If you have just overspent recently and have built up loans through your credit cards, bank or building society overdraft and HP agreements then you must do something to pay them off.

Have a long hard look at your finances and pare them to a minimum for a few months. Don't go to the pub where rounds of drinks will clobber your wallet, cut out the cinema, choose the canteen instead of the local wine bar for lunch, refuse to buy any new clothes for a few weeks. These sorts of cutbacks will be all it takes to get you back on the straight and narrow if the debts are still small.

But be brutal with yourself. Four or six savage weeks are better than six months of half-hearted cutbacks, and the quicker you pay off your debts the smaller the interest you will pay.

Once you are back under control, try to work out what it was that caused you to overspend. Are your credit cards leading you to buy on the never-never? If so there is a simple solution. Cut them up and send them back to the credit card company.

Doing that won't prevent you from re-applying at a later date. The credit card companies understand that some people go through stages when a card doesn't suit them. Providing you coped with the debt and sent back the card, they will be delighted for you to re-apply when you feel able to try again.

If you overspend because you enjoy shopping then you will have to call on good old-fashioned willpower to help you out. Either don't go shopping so often or limit the money you take with you. Alternatively, take up a hobby. If you spend your lunch hours in the gym or the swimming-pool, you could save yourself £s as you shed your lbs.

Difficulty in coming to terms even with small debts means you need help before they become big debts. Try a Money Advice Centre (your town hall should be able to give you details) or your local Citizens' Advice Bureau.

Large debts

Major debt problems are much more difficult to deal with. Recent research suggests that people deeply in debt respond to their problem with:

- secrecy
- piecemeal paying
- fear.

Secrecy. Partly because of embarrassment and fear of stigma but also because they believe that the more their creditors know about how bad things are, the nastier they will be.

Piecemeal paying. Paying off part of one debt, then a little bit of another because debtors have no idea that there is a possibility of dealing with all their creditors at once.

Fear. Worry over creditors' powers to imprison, take away your home and so on and fear of the courts, which are all viewed as being like a magistrates' court dealing with criminals. These worries are more often than not completely misplaced.

If you have serious financial problems, cutting back on your spending won't be enough. Most people with large debts have nothing to cut back on – the reason they have such huge debts is because they have no money in the first place. And it is not so much the size of the debt, but the ability to pay that counts. A rich man can cope with a debt of thousands or even millions. But a family on a tight budget could be sunk over a £100 loan.

However, there is a well-worn path back into the world of sanity, where the debt will no longer overwhelm you.

If you are in dire financial straits, take that path now.

- Face up to your problem. You won't sort anything out by refusing to admit you are seriously in debt. Debt is a four-letter word – but don't avoid saying it. If you cannot meet your financial obligations for three successive months you are in debt. Understand that and you are halfway to solving your problem.
- Work out how serious things are. Balance your books and see if your money problem can be solved by serious belt-tightening for a month or two. If you have gone beyond that then follow the next set of guidelines.
- Stop borrowing now. First, limit the damage, by cutting up all your credit and charge cards and don't sign any more HP agreements.
- Essential debts. There are some debts you must pay and these should be put at the top of the list. Essential debts are:

  Housing: rent, council tax, mortgage, and mortgage secured loans

  Service charges: water, fuel, phone

  Court orders: fines, maintenance, orders to pay creditors (these can all be varied by the court)

  Others: arrears of income tax and National Insurance, HP or other agreements where you do not own the goods.
- Don't rob Peter to pay Paul. Paying the debts that cause you the most hassle is not the answer, neither is trying to send a little money some months to appease all the creditors. You must take more serious action now.
- Draw up a personal balance sheet like the one opposite. Fill in all your income on the one side and all the necessary expenditure on the other. List all your debts.

Example of a personal balance sheet

Income		
Wages	£115	a week
Wife's wage	£80	
Child Benefit	£17.75	
Total	£212.75	
Spending		
Rent	£62	
Heating	£20	
Food	£40	
House insurance	£9	
School meals	£12	
Fares	£9	
TV licence and rental	£8	
Launderette	£3	
Pocket money	£1.50	
Total	£164.50	
Extras	£5.50	
Total	£170.00	
Surplus	£42.75	a week
Debts		
In-store card	£9.50	
Bank loan	£33.00	
HP	£16.00	
Total	£58.50	
Proposal		
New payments		
In-store card (£9.50)	£6.50	
Bank loan (£33.00)	£22.00	
HP (£16.00)	£11.50	

With luck your income will be larger than your expenditure – but remember to allow yourself a little money for clothes, occasional nights out and children's treats.

Now the income that is left over is the money that you must use to pay off your debts. Do it on a *pro rata* basis. That means you should pay more every month to the creditors to whom you owe the most and less to the creditors to whom you owe the least.

Don't overcommit yourself – it is important that you keep to the new schedule.

Now write to all your creditors. Explain to them that you have got yourself into debt but that you are trying to do something about it. Enclose a copy of your personal balance sheet showing how your finances add up and tell them how much you think you can afford to pay back every month. Ask if they will stop charging you interest otherwise your capital payments may make no inroads on the debt. If the debt is very high you could find that all your repayments are being eaten up just covering the interest, so you will not be doing anything to help yourself – the debt will not be being reduced. Not all creditors will do this, but some might and that would be a considerable help.

A simple but detailed balance sheet like the one above is very useful. It provides the creditor with the information he or she has a right to expect so that a decision can be made on whether or not to accept your offer. What the creditor will do is compare what you are offering with the likely outcome of legal action. If the offer is similar, they will probably accept.

If the matter went to a County Court judgement it would be unlikely that the creditor would continue to get interest on the money owed; that is why your creditor will often accept your proposal to stop the interest continuing to run on a private deal. The creditor will also want your repayments to be reasonable and realistic. Nobody wins if the payments break down.

Tip

Don't miss a monthly payment. Most creditors will accept this system of reduced payments, but if you miss a payment they will come down on you very heavily. That is why it is so important not to overcommit yourself at the beginning.

If you don't feel able to make out your own personal balance sheet and write to all your creditors yourself, ask for help from the Citizens' Advice Bureau or a local Money Advice Centre (your town hall will have details). They are used to dealing with this sort of crisis and a letter to the creditors from a professional CAB or Money Advice counsellor often carries more weight than one that you write yourself.

However, if you want to do it yourself you could try drafting a letter along the following lines:

account number — your address

date

Dear Sir,

At the end of this month I am being made redundant. Because I have only been with the firm for a year I will receive no redundancy payment. Although my wife is still working, and will try to do as much overtime as she can we will have to survive on a very much reduced income. After paying for all the household essentials there is very little left over at the end of the month.

Under these circumstances I will have to reduce the amount I repay to you from £33 a week to £22. I have enclosed a copy of my personal balance sheet and, as you can see, I should be able to cope with the new repayments. However, unless you can consider suspending interest charges on the debt, it will only continue to mount and my monthly payments will not make any inroads at all into the capital sum.

I am writing to my other creditors asking them to accept a similar arrangement and, in the meantime, I will be trying to find another job.

Although I am over fifty I am a skilled toolmaker so I hope to work again.

I shall keep you informed of all developments.

Yours faithfully,

Other steps worth taking

Maximize your income. Check that you are not missing out on any money that is yours by right. Your local CAB will be able to help you to check whether or not you are entitled to any DSS help – but make sure also that the Inland Revenue is not collecting too much tax. Does your tax man know that you are married? If not, you could be missing out on tax-free allowances.

Could you increase your income by taking a small part-time job, working from home, having a lodger. If your financial problem is serious you might have to think about one of these options.

Tip
You can now take in a lodger or lodgers and earn up to £62.50 a week tax-free. It could be your salvation.

Try to get help from other sources. In cases of extreme hardship your employer might help through a staff welfare fund or just through kindness, or you may be able to get some help from a charity. Ask your local CAB for any charity funds available.

Minimize your outgoings. By the time you get to the stage of serious debt, you have probably cut back on everything you possibly can. The odd half of lager in the pub is probably worth more to you in terms of mixing with other people than it costs you in income. But do just check through again in case there is something that you could trim back.

COURT PROCEEDINGS

There are several different ways in which a court can be used to sort out your financial difficulties. The legal systems of Scotland and England and Wales are not compatible – what you get depends on where you live.

Administration Orders (England and Wales only)

These orders can be used if you want legal help to clear your debts. The courts won't actually put up any money but they will 'administrate' the debt (hence the name) and keep your creditors off your back.

To qualify you have to:

- have more than one debt
- have total debts of more than £5,000
- have a county court judgement already against you.

You can follow the procedure yourself or ask the CAB to help you. First you need to fill in Form N92, An Administration Order Application. You can get it from a county court or from the CAB – they often keep a stock of them.

Take the form home to deal with it. It is not difficult but it can be time-consuming because you have to fill in all the details about your weekly incomings and outgoings, and list your debts. If you already have a personal balance sheet drawn up you can get the details you need from that. And be honest – there is no point in doing all this unless you are going to do it properly.

Take it back to the court to 'swear it in'. That means reading words from a card to swear that all the details on the form are true. You might also be asked for proof that the debts are true, so take along any bank statements, letters from creditors or credit card statements that would verify the information.

The court will then write to all the creditors that you have listed asking them to check the details and that should take about six weeks. After that a date will be set for a court hearing.

Don't expect it to look like a set from *LA Law*. The district judge hears the case 'in chambers' – that means you'll be in a small room rather than open court and a few of your creditors may turn up. The district judge will look at the order you have filled in, take into account anything the creditors say and what you say, and make an order for you to pay back a certain amount every month. At this point the interest on the debts will be stopped – and this is crucial because it gives you a chance to pay off the outstanding balance. Getting the interest stopped is the main reason people go for Administration Orders.

The amount set can be very low – perhaps £4 a month if you are unemployed. You pay the money into the court and the court divides it up among your creditors.

If you can't pay – tell the court immediately and they may set a new level if there is a proper reason (perhaps you have lost your job or your wife has lost hers). But if you don't pay because you have spent the money on something else you will be in breach of a court order and that is serious.

If you have built up a debt so large that you really have no hope of ever paying it back you could apply for 'composition'. This means that you are asking the court to set aside part of the debt and they may do it if, for example, you are a pensioner or a single parent. Not for nothing are Administration Orders known as 'poor man's bankruptcy'. The real drawback to an Administration Order is that you cannot get any more credit while you are paying off the debt – though this may be an advantage if you are not good at controlling your spending.

If your debts are larger than £5,000 you don't qualify for an Administration Order, but if you don't pay then your creditors will take you to court and much the same thing will happen. The interest will be stopped and a payments system set up.

Bankruptcy/Sequestration

If your debts have piled up to a level where you can never pay them back you may have to file for bankruptcy – or sequestration as it is known in Scotland.

This is not a step to be taken lightly. You may lose some of your debts, but you lose everything else as well.

Bankruptcy. In England you are charged for filing for bankruptcy – about £130. And bankruptcy is not automatically given. The Receiver could instead make you go for an Individual Voluntary Arrangement if you have assets of over £2,000. That means that you would have to sell your assets and pay the residue on a *pro rata* basis to your creditors. Mortgage, rent, tax and VAT arrears are not written off by bankruptcy, and even after you are discharged you cannot apply for any credit without disclosing that you are a discharged bankrupt.

If you decide to take this step, get hold of an Insolvency Service booklet called *Insolvency Act 1986 – A Guide to Bankruptcy Law* from 2–14 Bunhill Row, London EC1Y 8LL.

Sequestration. In Scotland, if you have debts of over £750 you can go to a solicitor or a chartered accountant and sign a trust deed which sets the sequestration process in motion. A trustee will then take control of your financial affairs and in the meantime you can't apply for credit of over £250 without disclosing that you are an undischarged bankrupt.

Warning

This is no easy step to get rid of all your debts in one fell swoop. It will cost you your home, any assets you have got and often your marriage.

Although it will take the immediate pressure off your financial circumstances, it will wreck your money plans for the rest of your life. You will probably never again get a mortgage, a bank loan or anything on HP. Credit and store cards will treat you like a leper.

If you are still quite young, it is a financial life sentence. Before taking any step in this direction get some professional money counselling – the CAB or a Money Advice Centre will be able to help you.

25 Crisis Times

Separation and divorce · Becoming sick or disabled · Widowhood · Redundancy and unemployment

Dealing with life's general ups and downs is difficult enough; dealing with the crisis times is almost impossible. Whether it is divorce, redundancy, widowhood, or serious illness, we are being confronted with emotional and financial problems we have never met before – at a time when we are least able to make rational and long-term decisions.

It is a time to get help from professional advisers – from people who are used to dealing with the crisis that has hit you for the first time and who know all the pitfalls. But you must still try and come to terms with the basic problems and have some understanding of the options open to you, or you won't be able to make the fullest use of their knowledge.

CRISIS AND DEBT

Because you are going through a trauma, money and money problems often take a back seat. If you have just lost your job, or recently been widowed, the last thing you think about is paying the mortgage or coping with the HP. But don't expect them to take care of themselves.

So whatever the crisis you are going through, try to take time to write to all the people to whom you owe money and alert them to your problem. Most will give you a sympathetic hearing even if your letter is simply a plea for three months' grace to allow you to get back on your emotional feet before you start coping with the money problems.

Very few people start out with any intention of getting into serious debt. However, a large mortgage, a few HP agreements, a bank loan for the car and a couple of credit card bills soon move from being manageable to unmanageable if one income is lost. So you have to move fast to avoid going under.

If you really feel that you are not going to be in a position to cope with even starting to sort things out, go along to your local Citizens' Advice Bureau for help. They have counsellors experienced in dealing with just this sort of crisis and will help you, confidentially, to take the necessary steps.

SEPARATION AND DIVORCE

Separation and divorce can seriously damage your wealth. Unless the couple are both young, earning and childless, or very rich, ahead of them lie years of financial hardship. Two may not be able to live as cheaply as one but, once a couple split up, they will soon realize that two living apart live much more expensively than two living together.

The actual cost of the divorce itself is not high. The petition costs around £50. But on top of that there will be legal fees to the two sets of solicitors involved in the case, and if you sell the house there are estate agents, removal men and lawyers to pay. If there are young children in the family, all this comes at a time when the finances are likely to be stretched anyway.

When you and your spouse separate or divorce, the first thing you should do is find a good lawyer. Don't pick one at random, nor necessarily use the one that did the conveyancing on your house. You will need a lawyer who specializes in family law. Ask any friends or colleagues who have recently been divorced who they used, or contact the Solicitors Family Law Association, 154 Fleet Street, London EC4A 2HZ. The association will send you a list of members in your area. Or try your local Citizens' Advice Bureau, which should also have a list of solicitors who deal with family law.

Your solicitor will explain to you exactly how the legal separation and divorce proceedings work and will help you to make the best of the financial situation and go for all the tax reliefs and DSS help you can. But solicitor's time is expensive – it can range from £50 to £200 an hour – so come to terms with the basic financial alternatives and take some of the necessary money steps before you go to see him.

What to do now

- Immediately you split up, tell the bank or building society from whom you have your mortgage. If the husband has moved out, but agreed to keep paying the mortgage, it is important that you know immediately a payment is missed. If the debt is allowed to mount it could become overwhelming. If you know the mortgage is being paid, don't tell the lender yet – just keep a monthly check on the payments.

- Don't go out on a credit card spending spree and assume someone else will pay. They won't.
- Register your interest in the home if it is entirely in your spouse's name, to stop your partner selling it without your agreement or using it as collateral for a loan. To do this make sure your solicitor puts a notice on the District Land Registry.
- Tell the DSS when you split up and get advice on benefits you might be entitled to, such as Income Support, Family Credit or Housing Benefit.
- Close any joint bank or building society accounts you have, or insist that withdrawals need both signatures. One partner could otherwise withdraw all the money or run up a huge overdraft and the bank or building society would be within its rights to transfer money from any other account you have to pay off the debt. Watch out, too, if you have joint credit cards. If your separation is still amicable this may be seen as a hostile move, so it may not be in your best interest to do this. However keep a close eye on things so that nothing gets out of hand.
- Write to the Inland Revenue if you are due an increased tax allowance such as the Additional Personal Allowance if you have children.
- Tell the local authority so that you stop being liable for your spouse's council charge.
- Try to get a half-hour fixed-fee interview with a lawyer who specializes in family law. Make sure you know before you go along what you want to ask, because half an hour is not long. At the end of the interview you should know what your next steps should be.

Scottish law

The legal side of divorce is different north of the border. Under Scottish law the court will try to adhere to the 'clean break' principle. That means they would rather give a lump sum to the have-not partner than running maintenance. So a wife might get a £20,000 payment, rather than £2,000 a year for ten years.

Pension rights will also be taken into consideration in Scotland. Life insurance and pension policies that could pay out large lump sums in the future will be considered and an award made so that one partner does not miss out.

Children's welfare will never be considered under the 'clean break' system. Maintenance will always be an obligation on both parents.

Separation and tax

While a couple are married they will be entitled to a personal allowance each, and a married couple's allowance.

In the year that they separate they will continue to have a personal allowance each, and they will still be entitled to the married couple's allowance.

If they have children who live with the wife, she can also claim the additional personal allowance if the husband is claiming the married couple's allowance.

In the following years, they claim what they are entitled to. That is, the husband and wife both have a personal allowance, no one gets the married couple's allowance and the parent with whom the children live gets the additional personal allowance. If each partner has a child living with them, they can both claim the additional personal allowance.

This system continues once the couple are formally divorced.

Divorce and maintenance payments

These can either be voluntary or enforceable.

If you pay maintenance *voluntarily* you can't be forced to pay up.

If payments are *enforced* under a court order, a separation deed or some other binding legal obligation, then they have to be made.

Court orders

(I am assuming the normal split up that the ex-husband pays maintenance to the ex-wife and children. If it is the woman who is paying the maintenance just substitute ex-wife for ex-husband in the text below.)

- The *ex-husband* will get full tax relief on payments of up to the level of the married couple's allowance per tax year on money going to the ex-wife.
- The *ex-wife* will pay no tax on maintenance payments, no matter how much is given.

- The *children* pay no tax on any money given to them, but the father gets no tax relief on the payments.

An ex-husband can give more than the married couple's allowance, but he won't get tax relief on it. Nor will he get tax relief on any money that is already free of tax – for example, mortgage payments that qualify for MIRAS.

Divorce and your home

From the date of your separation, or divorce, the tax man sees you as two single people again. That means you could each qualify for tax relief on the interest you pay on your mortgages of up to £30,000 each, a total of £60,000 a year. Of course the mortgages have to be on separate properties.

Tip

If the ex-wife is dependent on her ex-husband's income to pay the maintenance of herself and her children, she should take steps to insure his life. Buy term insurance, which pays a lump sum on death rather than a life insurance policy. It is much cheaper. (See p. 61.)

Divorce and pensions

From the date of the divorce, the wife loses all rights to the widow's benefits that are part of the husband's pension scheme. (It is unlikely, though not impossible, that the wife will have a widower's benefit in her pension scheme.) So she should ensure that some financial arrangement, such as a separate insurance policy, should be taken out to compensate her.

Divorce and the DSS

Avoid accepting maintenance payments which would take you marginally over the Income Support limit. A wife with no other income would be better on Income Support because she could also claim free prescriptions and dental treatment, free milk and vitamins for children under five, free school meals for the children, and – because the money is coming from the DSS – she will get regular payments. An ex-husband able to afford so little may be spasmodic about paying.

The wife may be able to claim special needs payments too, at times, from the Social Fund. Get help and advice from the Citizens' Advice Bureau or your local DSS office.

Divorce and your will

Marriage automatically revokes a will – divorce doesn't always. So make sure you rewrite your will after a divorce if you want to change the beneficiaries. This is obviously important if you are living with someone else and start another family. It is unfair to everyone you leave behind if you just expect them to sort it out after you go.

An ex-wife can bring a claim against the ex-husband's estate for financial support if she does it within six months of his death.

Cutting the cost of divorce

The cost of the divorce is, in the main, the cost of your lawyer. The more you use that professional the higher will be your bill at the end. I'm not suggesting that you cut out the lawyer, but here are a few tips to keep the bill as low as possible.

- Don't ring or write to the lawyer every time you have a question. Save them up and ask all the questions at one time. It'll be quicker.
- Try asking the solicitor's assistant. He or she will often know as much on the basic points of law and will charge a lot less.
- Don't use your solicitor as an emotional support. If you just want to talk, use a friend or relative who is not charging £90 an hour for listening time.
- Try and get legal aid. Although you may have to pay back the legal aid money once, say, a house or business is sold, it might tide you over for the moment.
- Have a round-table discussion – this means getting together with your solicitor opposite your spouse and theirs. With both lawyers present, tempers may not get so heated and you might be able to thrash out the basis of an agreement.

SICK OR DISABLED

At a time when you or one of your immediate family is going through a serious illness, or has recently become disabled, money worries are the last thing you need. Yet, inevitably, the illness itself brings with it an increase in financial needs. You might need money to pay for visits to hospital, to buy a lot of medicines, to fund increased heating bills if there is an invalid in the house all day, or to pay for a cleaner, housekeeper or nanny. And of course if it is the major breadwinner who has been struck down, you could find yourself without his or her wage just when you need it most.

Advance planning may reduce these worries. If you have taken out a permanent health policy remember to claim, and the weekly, monthly or lump sum benefit will help to tide you over.

If you are off work sick at least four days in a row, you should get Statutory Sick Pay from your employer or state Sickness Benefit for up to twenty-eight weeks.

Statutory Sick Pay. To qualify you have to work for an employer and pay National Insurance contributions. Depending on how much you earn, your SSP will be £45.30 or £52.50 a week. See p. 209 for full details.

Sickness Benefit. If you are self-employed, unemployed or can't get SSP from your employer, you might qualify for Sickness Benefit if you have paid enough in National Insurance contributions. SB amounts to £41.20 a week if you are under pension age, £51.95 if you are over pension age. See p. 209 for full details.

At the end of twenty-eight weeks your Sickness Benefit or Statutory Sick Pay ends. If you are still unable to work you must claim Invalidity Benefit of £54.15 a week instead. Invalidity Benefit is based on previous National Insurance contributions. You may also get Invalidity Allowance – for full details see p. 209. If you don't have enough to qualify, you may be able to claim Severe Disablement Allowance instead, at £32.55 a week. However, if you became disabled after your twentieth birthday, you must be severely disabled, that is at least 80 per cent disabled, to qualify for the allowance.

The real money worry about long-term illness is that unlike redundancy it doesn't come with a lump sum attached, so watch out for mounting debts. If there is no income to pay the mortgage don't just pretend that there is. Take all the necessary steps outlined in Chapter 14, Mortgages, to prevent yourself getting into serious debt. You should find that most building society and bank managers will be sympathetic, particularly if you go along early, before there is a money problem.

If you do recover from your illness and go back to work, your time in hospital and the financial crisis that accompanied it may encourage you to insure against any similar occurrence.

BECOMING A WIDOW

Sadly, becoming a widow can mean that a lot of the serious financial decisions that have to be taken, are taken in haste. Widows often spend the rest of their lives regretting the fact that they moved house, or paid off the mortgage, or invested their lump sum unwisely.

Of course certain steps have to be taken within hours and days of the death of your husband – the funeral has to be arranged, the death registered – but anything that can wait, should.

Years ago the death of a husband meant that, for the first time in their lives, women had to look after the finances of the household. Many felt they couldn't cope just because they were women. This is much less common now, but there are still a lot of new money decisions to be taken, decisions that have cropped up directly as a result of the death of the partner.

The funeral

If you have some savings of your own, you will get no help towards the cost of the funeral. In the late 1980s the average cost of burial or cremation was around £1,000.

Close relatives, the funeral director or your minister should be able to help you with the administration of the funeral.

If you don't have the money to pay for a funeral, ask for financial help from the DSS – but do it before you make any of the arrangements.

Do you need a solicitor?

To sort out your financial affairs and those of your husband you might need a solicitor. If your solicitor has been named as an executor of your husband's will then you will have his help automatically – though you will have to pay for it.

Otherwise it is up to you whether you do it yourself, or get help. If your affairs are relatively simple – that is the whole estate is being passed on to you, the bank and building society accounts tended to be in joint names and there were not a great number of other assets such as shares – then go ahead and do it yourself if you feel up to it. Free help is always available from the Citizens' Advice Bureau, who will be able to offer the basic guidelines.

If the estate is worth more than the inheritance tax level of £150,000 and some of the money is being passed on to other people, you might be well advised to get financial help. Of course, the more you do yourself, the less you have to pay to the professional who is helping you.

Mortgage

Whose name?

Go along and see the bank or building society manager who gave you your mortgage. If the home and the mortgage are in joint names, they will automatically belong to you. If they were in your husband's name alone, then make arrangements to have them transferred to you.

Pay it off?

Many widows immediately feel they want to pay off the mortgage. If they have any lump sum coming to them their first thought is that the money should be used to rid them of the monthly debt. But think again. A mortgage is the cheapest form of borrowing there is because you get tax relief on the interest. Once you pay it off it is impossible to take it out again, and get tax relief, without moving home. So, should you need the money at a later date, even if it is just to go on a well-deserved holiday or a visit to Australia to see the grandchildren, you will have to take out a loan at a much higher rate of interest.

If there was a mortgage protection policy or an endowment plan on your husband's life, the mortgage will be paid off automatically.

Can't pay?

If you feel you cannot afford the payments, you may be able to get help elsewhere. First you must work out whether your problem is temporary or permanent.

Temporary. If you are going to have difficulty for a few months while you sort out your husband's affairs and release some of the capital, then your bank or building society will probably agree to your not paying the monthly repayments for up to three months.

Permanent. If you feel that the payments are too high for your income, and are going to continue that way, then you must choose a different course. You could extend the life of the mortgage – that way you would pay less each month, though for a longer period. Or you could try to get help from the DSS. They will sometimes make the interest payments on the mortgages of people who cannot pay them themselves. However, if you think you are never going to be able to pay off the mortgage you must reassess your future and consider selling your home and buying something smaller.

There are ways of making your home pay for itself – either by taking in lodgers or (if you are old enough) using a home-income plan or reversion policy to unlock the capital of the house. See Chapter 28, Housing for the Elderly.

Renting

If you rent your home, and the tenancy was in your husband's name, change it into your name. You can do this providing you were living in the property immediately before your husband died.

Don't worry about being evicted. A landlord cannot force you to move out unless you have stopped paying the rent or in some other way broken the terms of the tenancy.

Tax

In the year that the husband dies

- The married couple's allowance and his personal allowance will be offset against his income.
- The wife's personal allowance will be offset against her income.

She will also get a widow's bereavement allowance, which is equal in value to the married couple's allowance. If she gets all or part of the married couple's allowance this will eventually be stopped – she cannot get both.

She will also get an additional personal allowance if she has any children.

In the year after the husband dies

- The widow gets the personal allowance and the bereavement allowance and, if she qualifies, an additional personal allowance.

There is no widower's bereavement allowance.

If you have children who are still in full-time education or on a two-or-more-year training course, you can claim the additional personal allowance for them (but it is only one allowance no matter how many children there are).

There is a helpful booklet available from the Inland Revenue: IR91, 'Independent Taxation – A Guide for Widows and Widowers'.

Widow's benefits

There are three types of benefit that a widow could be entitled to:

- Widow's Payment
- Widowed Mother's Allowance
- Widow's Pension.

They are all dependent on the husband's, not the widow's, National Insurance contribution record.

Widow's Payment. This is a tax-free cash payment of £1,000.

The widow will not qualify if:

- she is over sixty *and* her husband was receiving a state retirement pension when he died
- she is divorced
- she was never married to the man
- she is living with another man as if she were married to him.

Widowed Mother's Allowance. This is a weekly payment of up to £54.15 a week, plus £10.85 a week for each of the children. The widow may also get an Additional Pension based on her husband's earnings since 1978.

To qualify, the widow or her husband must have been receiving Child Benefit or she must be expecting her husband's baby.

Widowed Mother's Allowance is taxable and is paid as long as she is claiming Child Benefit.

Widow's Pension. This is a weekly payment of up to £54.15 a week. The widow may also qualify for an Additional Pension based on her husband's earnings since 1978.

To qualify, the widow must be at least forty-five when her husband died or when her Widowed Mother's Allowance ends. You cannot qualify for both the Widowed Mother's Allowance and the Widow's Pension.

The Widow's Pension is taxable and will be paid until the woman is entitled to a State Retirement Pension.

To claim

Fill in Form BW1 to claim for all the benefits and do it within a year of your husband's death if you are claiming for yourself, within six months of your husband's death if you are claiming for a child as well.

Further information in DSS leaflets FB29, 'Help When Someone Dies', NP45, 'A Guide to Widow's Benefits', D49, 'What to Do after Death' and NI51, 'National Insurance for Widows'.

If you can't cope with all these forms and arrangements at a time of deep emotional distress, ask the Citizens' Advice Bureau for help.

Pensions

If your husband was in a company pension scheme, or had made his own pension arrangements, you must write to the pension company asking what your benefits are likely to be.

In a company scheme the benefits could come in two forms:

- a lump sum
- an annual or monthly pension payable to the widow.

Neither is compulsory; some companies do both and some do either.

Lump sum. This will be related to annual pay and can be up to a maximum of four times annual pay. It is more likely to be twice the salary.

Pension. This will normally either be a multiple of annual salary, or related to the pension the husband would have been paid.

The payment of the lump sum is at the discretion of the trustees – this means that it is up to them who gets it. In most cases it will be paid to the widow, but it is not hers of right.

The pension goes to the widow as of right – no matter how long she and her husband may have been separated, or how long he has lived with another woman.

DSS

If you find after your husband's death that you do not have enough money to live on, knock on the door of the DSS. There is quite a lot of help you can get there in the way of Income Support if you have a low income, Housing Benefit if you cannot afford your rent and council tax, Family Credit if you have children and work more than twenty hours a week, and in some cases special loans for particular needs from the Social Fund.

If you have difficulty with large payments such as fuel bills or insurance try to spread them over the year. See Chapter 24, Coping with Debt.

REDUNDANCY

Even if you are vaguely expecting it, a redundancy notice will hit you like a bombshell. In a few weeks there will be no job for you to go to, and more importantly, no wage coming in.

Like all the other crises, the emotional upheaval will be as debilitating as the financial one, but at least you have one positive advantage. If you worked for your employer for more than two years you will get some sort of redundancy pay-off.

What is redundancy?

Redundancy means that your job has disappeared. Perhaps your employer has gone into liquidation, you've been overtaken by technology, your job is moved to a different part of the country or there is just a downturn in business – whatever it is you are surplus to requirements.

If you are dismissed and someone else replaces you, that is not redundancy, that is good old-fashioned 'sacking', and if you feel it is unfair try taking your case to an Industrial Tribunal on the grounds of unfair dismissal.

You cannot be made redundant if you are under eighteen, over the age when you can start claiming the state pension or work for your spouse. Redundancy doesn't affect dock workers, crown servants, employees in a public office or in the NHS, as they all work under a different system.

Redundancy payments

You are entitled to:

- *18–21-year-olds*: half a week's pay for each full year worked

- *22–41-year-olds*: one week's pay for each full year worked
- *41–60/65-year-olds*: one and a half weeks' pay for each full year worked up to a maximum of thirty years.

And there is an earnings limit. No matter how much you actually earn, there is a cut-off point, currently £220.

So a woman who worked for her firm for thirty-two years earning £220 a week when she was made redundant would receive a lump sum of:

30 × £198 × 1½ = £8,910.

That is the minimum a firm is forced to pay by law. You may be lucky and find that your firm pays more – particularly if they are looking for voluntary redundancies and are offering good severance pay to encourage people to go.

Redundancy pay and tax

Redundancy pay is not taxed . . . unless:

- your employer is offering more than the statutory minimum because your contract of employment lays down that more must be paid
- you are being offered a 'golden handshake', that is a financial sweetener to encourage you to take early retirement or to compensate you for job loss.

Golden handshakes are tax-free up to £30,000 and are then taxed at your top rate. If you are likely to be unemployed for longer than a month write to your local tax office in case you are due a tax rebate.

Redundancy and the bankrupt employer

If your employer has gone bust it is unlikely that there will be any money in the kitty for redundancy pay. Your money in lieu of notice may not even be there. If that is the case, ask the receiver or liquidator who is now running the company for a leaflet entitled 'Employee's Rights on Insolvency of Employer'. It explains how to apply to the government's redundancy fund for your money.

Redundancy and your pension

Your pension scheme often doesn't realize you are being made redundant, it just sees you as no longer working for the firm and therefore not contributing to the pension fund. So your choices on leaving due to redundancy are the same as at any other time.

You can leave your pension in the scheme, transfer it to a new employer, or buy a personal pension plan. See Chapter 27, Pensions, for fuller details.

Investing your lump sum

A redundancy payment is a marvellous windfall. At a time when you feel you are being thrown aside and you question whether you will ever work again, it is nice to have a bit of money in your hand. For many people it is the first time they will ever have such a large sum of money and there is a great temptation to spend, spend, spend. A holiday, a car and a new kitchen will make quite a hole in anyone's redundancy payment so don't be tempted to blow the lot.

Your redundancy pay is meant to tide you over the bad times, till you find work again. Use it for that.

The first thing you will have to do is work out what you need from the money – income or capital growth. Once you have done that, use Chapter 3 to find the best place for your money. Be wary of investing in anything risky. If your redundancy money is to be your main source of income over the coming months or years you cannot afford to take any risk at all with it.

Beware of the professionals. If your company is making a lot of people redundant in your area you'll find yourself being 'cold-called' by professional money advisers – that means they ring you up without being asked and try to sell you financial products. Some will be decent, honest and truthful, some won't, so treat their advice warily. However sympathetic they seem, they will all be trying to sell you something so that they get the commission. If you know the basics of investment yourself you will at least know if the advice is good or bad.

UNEMPLOYMENT

For most people redundancy is immediately followed by unemployment. Don't put off going to your local unemployment office. The benefits are yours by right – you paid for them when you were working – and they may be able to help you find a job again. Even if you don't want the benefit, register as unemployed so that you protect your National Insurance contributions record.

Benefits

Unemployment Benefit. You can claim this on the first day you are unemployed provided you have paid enough National Insurance contributions and you are willing to get another job.

A single person under retirement age currently gets £43.10 a week, and this is not a means-tested benefit. You get £54.15 if you are over retirement age. If you are married with a wife and family to support you should be able to get other DSS assistance – Income Support – to help with rent and council tax, running the household and supporting your family. You may be able to claim for your wife or husband under the Unemployment Benefit scheme. Providing they earn less than the benefit you could get an additional £26.60 a week.

Unemployment Benefit is taxable. That does not mean that the DSS will deduct basic rate income tax before they pay your money. It means that if your income in a tax year is more than the personal (or married couple's) tax-free allowance then you will have to pay tax on the money.

In practice for many people it means that if they lose their job midway through a tax year, they won't get a tax refund, or the refund will be affected by the unemployment benefit and unless you have substantial untaxed income you would be unlikely to get any sort of tax bill.

Tip
Unemployment Benefit is never backdated, so move sharply down to the UB office. If you don't claim you won't get it and it is yours by right if you have made enough NI contributions.

Income Support. For many people unemployment benefit just is not enough to cover all their basic needs. If that is the case apply for Income Support.

This benefit is means-tested first on your savings, then by your income. If your savings are under £3,000 it will not affect your Income Support. If they are between £3,000 and £8,000 it might. Every £250 or part of £250 that you have above £3,000 will be seen as providing a weekly income of £1. If you have savings of over £8,000 you cannot qualify for Income Support.

Housing Benefit. Works in a similar way to Income Support, but up to a ceiling of £16,000.

How much benefit you get depends on how much money you have coming in every week and how much you need to live on and that will be calculated for you by the DSS, using rates laid down by Parliament. It is likely to have a different view from you on how much money you actually *need* every week.

If you are on Income Support you will get help with part of your rent and poll tax/council tax (but not your water rates) as well, and perhaps your mortgage interest too.

There are other benefits, too, such as free school meals, milk and vitamins for your children, free prescriptions, help with dental treatment, and NHS glasses. What you get depends on your financial circumstances, and if you want more help ask your local Citizens' Advice Bureau or use the DSS freephone help-line, tel. 0800 666 555. Or get a copy of the DSS leaflet FB2, 'Which Benefit?'.

Unemployment and your mortgage

Once you lose your job, you may not be able to keep up the mortgage payments on your house. Don't just stop paying them. That will mean that your arrears will start to build up. Tell the lender what has happened – that is a good first step. Then try to get the DSS to pay the interest on the mortgage. The DSS may pay half the interest on the mortgage for the first sixteen weeks that you are unemployed and then all the interest. That is usually acceptable to the lender if they know what is happening. But the DSS will not make capital payments to reduce the lump sum, nor will it pay the premiums on your endowment. These you have to make yourself.

If you are not going to be able to manage with this help, do something about it now. Get professional counselling from the CAB or from your lender, and read Chapter 24, Coping with Debt.

26 Retirement

Getting ready for retirement – the steps to take on pensions · Investments · Debts · Age allowance – how to avoid the trap · What happens if you are too old to manage your money

See also

- *Chapter 27, Pensions*

PLANNING FOR RETIREMENT

Most people plan for their retirement in the few years before they actually stop working.

In their twenties, starting out in life, there's too much to do buying their first home.

In their thirties it is raising a young family that takes all the money.

In their forties it is coping with the problems of teenagers, exams and their children getting a job.

So for many people it is not until they are in their fifties that they begin to look ahead to retirement and making sure that the money, at least, will be ready for retirement when that day comes.

Expectations of retirement are much higher nowadays than they were a generation or so ago. In the past, workers often dreaded retirement because of the subsequent loss of earnings. They expected to have difficulty making ends meet.

But now, with personal pension schemes and the government-backed state retirement pension, many people look forward to their retirement. They may accept that they will have to trade down a bit with the car and perhaps even the house but they don't want to start taking the bus or moving to an area or a flat that they wouldn't like. Most people look forward to retiring because of the extra time they will have to do all the things they've never managed, visit all the places they have always wanted to go – and they certainly want to have the money with which to do it.

To make sure that you are going to get the best out of your retirement there are several areas you should check while you are still earning and able to make financial provision if you have to.

Your pension

Your main source of income once you retire will be your pension. If you have worked all your life it should be made up in two parts – the state retirement pension (the old age pension) and the state earnings-related top-up. You might also have a private pension.

What you want to know is how much you will get every month. Assuming you are entitled to the state pension you can ask the DSS for the current rate.

On top of that you may qualify for a SERPS payment – that is the State Earnings-Related Pension Scheme. Ask the DSS, using form NP38, what your entitlement is likely to be.

Then find out how much your company pension, or any private plan you have taken out yourself, will be worth. Although you won't be able to get a completely accurate figure – partly because most pensions are calculated on final salary or the final investment value of the money paid in, and you don't know what that will be – you should have enough of an idea to know whether or not there will be enough money coming in for you to live on.

This is only a brief outline of what pensions are and how they work, full details of all sorts of pensions are in the next chapter, so if you want more information now move on to p. 315.

If you are not going to have enough to live on, do something about it now. The earlier you start, the more you will get. You have two main choices as to what you can do:

- increase the amount of money you pay into your pension scheme
- save as much as you can so that you will have a large capital sum to invest for income.

AVCs

Paying as much as possible into your pension scheme is the most tax-efficient way of helping your retirement. Additional voluntary contributions – AVCs as they are known – are well established. It is the official title for the extra money you pay into your company scheme on top of the sum you are contractually obliged to pay. For full details see p. 326.

The drawback of paying AVCs is that you lock up your money. The only way to get it back is to retire. A drastic step if you only want a thousand or so pounds out for a holiday.

So don't over-invest for your retirement; keep some of your savings in an accessible building society or bank account or PEP.

Investment

As you approach retirement, your investment priorities change. The decade before you stop working you probably have a bit of extra cash around that you haven't had before. It is spare cash that you may have been willing to take a bit of a risk with – buying shares, investing in unit or investment trusts, maybe even going as far as gambling on the traded options market or risking a few hundred or a few thousand pounds on speculative shares.

If that is the case you will no doubt have had a few winners and a few losers.

As you approach and enter into your retirement, you'll find that, psychologically, you don't want the losers. That means of course that you will not then be able to risk any of your cash because it would be a fortunate man indeed who only picked winners.

Safety first will become your motto. Your lump sum or savings will tend to gravitate towards building society and bank accounts, gilts, National Savings, a TESSA and perhaps a few solid unit or investment trusts or a Personal Equity Plan. For full details on how these investment vehicles work, see Chapters 2, 3 and 5.

Start laying the ground rules now. Work out how much, if any, of your savings you would want to risk. And rationalize your finances accordingly. You don't need sixteen building society and bank accounts. Keep your affairs simple and they will be easier for you to look after – and easier for anyone else to cope with should something happen to you.

Paying off debts

Psychologically it is nice to step into retirement with a clean sheet – all your debts paid off.

It doesn't always make good sense. If you have taken out a loan you may be charged a penalty for paying it off early. And this is one place where the Consumer Credit Act has unwittingly acted against the consumers' interest.

In the past, if you paid off a bank loan early, the bank would refund to you any extra interest you had paid. Then came the Consumer Credit Act saying that a lender was entitled to take the equivalent of two months' interest – but no more – to cover administration and other expenses if a loan was paid off early. So the banks said thank you very much, and they may now charge you two months' interest if you repay your loan early. Of course, HP agreements and any other form of loan – which in the past may have charged much more than two months' interest for early repayment – are also now limited.

When it comes to your mortgage, follow this guide:

Repayment mortgage. If you are in the last four years of its term, pay it off if you can. By the time you reach retirement age the amount you owe will be quite small so you won't be paying much interest on the lump sum. Since it is only the interest you get tax relief on, you won't be getting much tax relief. You might as well get rid of the debt if you prefer not to have it.

But, before you do anything, ask your bank or building society manager for details of early repayments. Sometimes you have to give up to three months' notice (or face a penalty charge equal to three months' payments). So give the notice. If you pay the loan off early, you should be released from the interest you would have paid.

Tip

It is worth remembering that once you have paid off your mortgage, it is difficult to take it out again without moving house. So don't use all of your savings to pay off the loan if your pension will cover the payments. Because of the tax relief, a mortgage is the cheapest form of borrowing there is, so definitely don't take out any other sort of loan to pay off the mortgage.

Endowment mortgage. Keep it on until the end of the term. It is in the last few years that an endowment policy really comes into its own and earns big money, so by continuing with the payments you could build yourself up a nice lump sum over and above what you need to pay off the mortgage. If you won't be able to afford to make the payments after you retire, turn to Chapter 14, Mortgages, to find out what your options are.

Planning ahead

Finally, think ahead to what life is going to be like when you retire. What you are going to need, what you may need to buy. Funding the large purchases will be much easier when you are still working than when you are starting to cope with a reduced income for the first time.

For example, will you need a car? If you have always had a company car will you be able to buy it when you retire? Or will you want to sell the car you own and buy a smaller one when you retire?

Does the washing-machine need replacing? Is there any major work needing to be done on the house? Are you planning a holiday of a lifetime? If you are going to make a major purchase in the first few weeks of your retirement, try to have the money already saved before you give up work. Don't rely on the lump sum from your pension fund to put in a new kitchen, buy a car, or pay for a QE2 cruise – that money will be needed elsewhere. After all, you do have to live and eat in the golden years of your life.

RETIREMENT

The problem with living is that we don't know anything about dying – particularly when it is going to happen. If we knew that it would make arranging our money affairs much easier.

If I knew I was going to die next year, I'd spend, spend, spend now and certainly put nothing into a pension!

It is particularly troublesome, in money terms, for pensioners. They don't know when they are going to die – so they can't apportion their money.

Most people fall into one of two financial categories: either you believe money was made round to go round or you believe it was made flat to pile up.

If you are in the first category you will probably spend your way through any lump sum you get on retirement in the first few years. Whether it is a new car or kitchen, a holiday of a lifetime or just a good time in the pub, you will have enjoyed spending it – but it will be gone.

Alternatively you will save it and any other spare money you have because you worry about living to be a ripe old age on a reduced and fixed income and running out of money. The chances are you will die rich.

The trick – when you become a pensioner – is to do a very clever balancing act: spend enough to enjoy your retirement. But save enough so that you don't run out of cash.

Retirement certainly affects your finances – how much and in what way it affects them depends on your age when you retire.

Your main source of income will probably be your pension. The more you have paid into pension schemes, and the earlier you started paying, the more you will get out in retirement. For full details of how well off you should be, turn to the next chapter.

Investing your money

Most people take a large proportion of their pension as a lump sum and, if you have some savings too, you have to start thinking about what you should be doing with your money.

First work out whether what you want from your money is:

- income

or

- capital growth.

As a rule of thumb, if you can't live on the money you have coming in every month, you will need income from your lump sum.

If you can live on the money you have coming in every month, you may want capital growth.

Now that you have given up work, and so have no regular and rising income, you will find that your outlook on investment changes. You won't want to take much risk with your money.

You will probably prefer to keep more in safe places like building society and bank accounts, and your investment in shares will probably be limited to unit and investment trusts, where any risk is well spread. Your tax rate should drop too, if you have been a high earner in the past.

Age allowance

There is one concept that you should understand at this stage, and that is age allowance.

Everyone whether they are single, married, divorced or widowed gets a tax-free allowance. That means that you can have an income every year on which you pay no tax.

But once you reach the age of sixty-five you get a larger tax-free allowance and it is known as the age allowance. It increases again at age seventy-five.

At the moment (tax year 1992–3) the tax-free age allowance is:

- personal allowance £4,200
- married couple's allowance £1,720
- personal allowance (75 and over) £4,370
- married couple's allowance (75 and over) £2,505.

Unfortunately the concept is nothing like as simple as that.

The age allowance is designed to help pensioners financially by giving them this extra tranche of tax-free income. The logical conclusion of that argument is that if a pensioner is well-off he shouldn't need that extra help so he won't get it. And that's precisely what happens in a complicated sort of way.

Everyone over sixty-five (man or woman) gets the age allowance. If you are well-off and don't deserve it, then it is taken back again. The Inland Revenue definition of well-off is having an income of over £14,200 in the 1992–3 tax year.

What happens is this: once your income breaches the £14,200 barrier then you start to lose some of your age allowance. The allowance is reduced by £1 for every £2 you earn over the basic age allowance. And instead of the age allowance you end up with the ordinary personal or married couple's tax-free allowance.

What to do?

- If your annual income is well below £14,200, you have nothing to worry about. You are not a taxpayer so make sure that no tax is being deducted from any savings account interest.
- If your annual income is well over £14,200, there is nothing much you can do – you can't stop what is happening.
- If your income is just below or just above £14,200, there is quite a lot you can and should do to avoid this swingeing penalty.

Every £2 in income that you get costs you £1 in tax – so avoid taking any income on that money if you can.

Invest it for capital growth if you don't need the income. If you make a capital gain on the money you won't pay tax at all unless your profits in any one year exceed the CGT limit (£5,800 in the 1992–3 tax year).

You could put your money into gilts – and a good ploy for pensioners caught in the age allowance trap is to buy a gilt due for redemption in a few years' time. The profit that you make will be CGT- and income-tax-free, and you are not tying your money up. You can always sell early if you need the cash.

If you really want income from your capital, move it instead to National Savings certificates. The interest is tax-free so it won't affect your age allowance. Of course the interest is added on to the certificates every year so to get an income you would have to redeem certificates. The interest is paid on a sliding scale – the longer you leave the money in, the higher the rate of interest you get. So don't use this route if you know that you will need the money within a year of buying the certificates.

Income tax

Provided you have alerted the Inland Revenue to the fact that you will be retiring the following year you should have no extra tax to pay, nor a tax rebate due, on your retirement.

But if you are going to continue to work after your retirement, perhaps part time or fewer days, make sure that your tax code is changed, otherwise you will pay too much tax.

Remember that, although retired, you are still a taxpayer. I get a lot of letters from pensioners assuming that, just because they have retired, they no longer have to pay income tax. Wrong.

If your income in any one tax year exceeds the tax-free allowance that you are entitled to, you will pay tax on the surplus no matter how old, or for that matter how young, you are.

Working in retirement

You no longer lose any of your state pension when you take a job. In the past, earnings of over £75 a week affected your pension, and the DSS docked money. This is no longer so.

You will, of course, still be liable for income tax if your income overtakes the tax-free allowance to which you are entitled.

Enduring Powers of Attorney

If you worry about getting so old that you cannot manage your financial affairs, there is something you can do about it now. You can have a solicitor draw up an Enduring Power of Attorney.

You are, in effect, giving someone power over your affairs should you become mentally incapacitated and unable to handle them yourself. It could be a relative or a professional such as a solicitor, financial adviser or bank.

The Enduring Power of Attorney has to be drawn up on a special form and signed by you while you are still 'of sound mind'. It is a new piece of legislation which enables the Power of Attorney to continue on through mental incapacity. In the past if you became mentally incompetent, the normal Power of Attorney would lapse and your relatives would have to go to court to get their powers back to look after your affairs. Under the new legislation, if you do become mentally incapable, the person you want to look after your affairs will apply for the power through the Court of Protection.

An Enduring Power of Attorney is a major step and I wouldn't recommend that you use an HMSO form without seeing a solicitor first. Once signed the EPA can only be revoked while you are still 'of sound mind'.

Concessions

Pensioners can usually make use of a lot of concessions. When you are working you tend to use money to save time; but once you retire get into the habit of using your time to save money.

Travelling off-peak is the simplest of the money-saving habits to get into – many of the others will be local to the area you live in.

Even if you don't fancy joining a pensioners' club or day centre, go along and have a look at the notice-board. It may have details of money-saving schemes.

Plenty of insurance companies offer cheaper deals for pensioners. If you are over fifty-five or sixty you can often get cheaper motor insurance – because the older you are the safer you are as a driver, say the statistics.

And you may get cheaper house insurance.

So the insurance company that offered you the best deal during your working life may not be the best one to stay with now. Check around.

27 Pensions

The state pension, your company pension, personal pensions – what they are, which is best for you · Retiring early, retiring late, changing jobs · What questions to ask a pensions' salesman

See also

- *Chapter 26, Retirement*

In children's story books there is often the tale of the little red hen who decided to bake a cake. None of the other animals in the farmyard would help her to gather the ingredients and mix them together, but they all wanted a slice of the cake once it was baked.

Pensions are a little like that. None of us really wants to contribute to the fund during our working life – but we all want our fair share of the pay-out when we retire.

Pensions work on the principle that the more you put in, and the earlier you start to fund your pension, the more you will get out on your retirement. Unfortunately, pensions are not quite as simple in practice as that theory implies. In the past some people, particularly job-changers, have missed out.

Far-reaching changes have been made to the UK pensions system in recent years – most of them for the better.

The new legislation has affected almost everyone who works, because it has opened up the available pensions choices. You can now contract in, or out, of SERPS (the earnings-related part of the state pension); you can stay in your employer's pension scheme (if he has one), or opt out into a personal pension of your own; and job-changers are now given more freedom as to what they can do with their pension when they move from one firm to another.

Pensions legislation was changed for a number of reasons:

- to encourage people to provide for their own retirement

- to give people more choice
- to reduce the financial burden of SERPS to future generations
- to increase the number of people in pension schemes
- to allow people to change jobs without reducing their pensions rights.

The new legislation was in place by July 1988 and the 1989 Finance Act tied up a few loose ends. Its most important effect is on attitudes. In the past, your company either had a pension scheme or it didn't. Very few people, working for an employer without a pension scheme, took out their own pension plan. Now, because you really do have a choice about what you do with your pension and where you want to save for your retirement, you have to do your homework.

So, you have to know how to choose the best deal for your financial circumstances, and you have to do it at various stages in your life.

Before you can choose, you have to understand how pensions and pension schemes work.

There are three types of pension:

- state pensions
- company pensions
- personal pensions.

THE STATE PENSION

The state pension is paid by the DSS to all qualifying individuals when they reach the age of sixty-five for men, or sixty for women. It is never paid before that age, though you can delay claiming it for up to five years and get more every week when you do claim. The pension is taxable, if you are a taxpayer.

It is divided into two parts:

- the basic pension
- SERPS.

The basic pension

The old age pension is a flat-rate pension paid to anyone who has paid, or been credited with, enough National Insurance contributions during their working life. That means you must have been paying for 90 per cent of the years between age sixteen and age sixty-five (for men), or sixty (for women) to get the full amount.

Men have to show a contribution record of at least forty-four years, women of thirty-nine years.

If you are working, you make regular contributions for National Insurance. They will be deducted automatically from your salary if you are an employee; you will have to pay them regularly if you are self-employed. However, if you are not working for any of the reasons below you will receive NI credits. That means that your NI record will not be broken during that period.

You will get credits if:

- you are claiming Unemployment, Maternity, Sickness or Invalidity Benefit, or Statutory Sick Pay
- you are a man aged between sixty and sixty-four and not in work
- you are still in full-time education
- you are on an approved training course.

The 1992–3 pension rate is:

- married man's basic pension £86.70 a week
- single person's basic pension £54.15 a week.

Of course if husband and wife both qualify for a full state pension in their own right because they have paid enough NI contributions, then that is what they'll get: a full state pension each of £54.15.

Gaps in your record

If you are at home looking after children and claiming Child Benefit, or looking after an elderly or sick person, you will receive credits for this period (unless you have received Invalid Care Allowance). This is known as Home Responsibilities Protection.

If you stopped working voluntarily, or have been working abroad, you might have gaps in your NI contributions record. You should fill these gaps to retain your right to a pension if you stopped work for more than a year or two. To do this, you pay what are known as Class 3 NI contributions – in 1992–3 that would be £5.25 a week.

If your gap was before 1982 you will have missed the boat, but after that date you have six years to make up the contributions. People nearing retirement could find that making up recent gaps will entitle them to a full, rather than a reduced pension. Write to your local DSS office, and they will tell you how much you owe.

SERPS

With the whole of the state pension system – the basic flat rate and SERPS (the state earnings-related benefit) – the money you pay in does not build up in a fund for you to draw on. It is used to pay the pensions of the current pensioners and other social security benefits. Your pension will be funded by the next generation of workers.

The worry is that there will not be enough workers to pay for the SERPS part of the pension. So the Conservative government used pensions legislation to reduce the benefits of SERPS and encourage people away from it.

SERPS was introduced in 1978 and is an additional pension, on top of the basic pay-out, which you will be entitled to if you have paid enough in National Insurance contributions. Each week that you earn more than the lower earnings limit (1992–3, £54), part of your National Insurance contribution goes towards your SERPS pension. The more you earn, the more goes into SERPS – up to an upper earnings limit (in 1992–3 of £405).

You won't be in SERPS if:

- you pay the reduced-rate married woman's stamp
- you are in a 'contracted-out' private pension scheme
- you are self-employed
- your earnings have always been less than the lower earnings limit.

How much will you get from SERPS?

In April 1988, the Conservative government moved the goalposts on SERPS. The new way of calculating how much you will get means that you get less if you are not retiring this century.

If you retire before 6 April 1999, the changes won't affect you; those retiring between April 1999 and April 2009 will be partly affected; those retiring after April 2009 will get much less than they might have envisaged.

Unless you have the brain of a first-class honours student in mathematics, you probably won't be able to work out your SERPS entitlement. Don't worry, the DSS will do it for you. Apply, using Form NP38, or phone the DSS freephone: 0800 666 555.

WHEN TO RETIRE

Retiring at sixty

State pension

If you are a woman and have enough full National Insurance contributions in your own right, you are entitled to a state pension at the beginning of the week following your sixtieth birthday. You won't get it unless you retire. If you haven't worked, or if you opted (while you still could) to pay the smaller contributions that married women could make, then you will have to wait until your husband becomes eligible for his pension before you get yours.

If you are a man you will have to wait until your sixty-fifth birthday – regardless of the age at which you retire. If you retire early, and don't take another job, you won't have to make any National Insurance contributions after the age of sixty. You will be given automatic credits. If you do work, and earn more than the lower earnings limit (£54 in 1992–3) then you will have to pay National Insurance.

Company pensions

Many company pension schemes continue to conform to the state retirement ages of sixty-five for a man and sixty for a woman. But many more are converging on a common retirement age. A European Court ruling, known as the Barber judgement, has outlawed sexual discrimination in pension schemes. This means that it could be illegal for men and women making the same pensions contributions to have different retirement ages.

And in most cases, the retirement age is equalling at sixty-five for everybody. That means that women, as well as men, will have to work to sixty-five in order to retire on their full pension. However, a woman wanting to retire at sixty (which would now mean taking early retirement) would be unlikely in the immediate future to miss out. Most schemes would continue to treat her generously.

Retiring at sixty-five

State pension

If you are a man you will be entitled to the basic state pension at the beginning of the week following your sixty-fifth birthday – providing you retire. You will get £54.15 a week if you are single, £86.70 if you are married, regardless of the age of your wife.

And you qualify for the basic state pension if you have paid National Insurance contributions for 90 per cent of your working life.

Company pensions

Most company schemes pay the full pension to men retiring at sixty-five.

Retiring early

State pension

The state pension is not paid until you reach state pension age, regardless of when you actually retire.

Company pensions

If you want to retire early, most company schemes will be able to make provision to pay you something.

How much you get depends on when you want to retire but most employees are surprised when they find out how much their pension would be reduced if they opt for early retirement.

A sixty-five-year-old man on £20,000 a year would expect to get a pension of £10,000 a year from his group pension on his retirement after thirty years with the firm.

If he opts to retire at sixty instead he could get the amount shown in the example below:

Reason for retirement	*Pension per year*
Voluntary	£5,800
Ill-health	£10,000
Redundancy	£9,166

(All pension schemes are different, the above figures are only an example.)

Some firms will give generous terms if you are retiring early through ill-health or if your employer is asking you to take early retirement, for whatever reason. If you are thinking about leaving early because you no longer want to work, find out how much your pension will be affected before you hand in your notice. As the table shows, you could be in for a nasty surprise.

The Inland Revenue will allow you to retire, for whatever reason, from the age of fifty – and that applies to men and women – but if you are opting for early retirement before that age you would have to show very serious ill-health before the Inland Revenue would allow it. Pensions are very tax efficient and the tax man wouldn't like to think you were using your pension to avoid paying tax on your earnings and then trying to do a clever tax dodge by retiring at thirty-two!

Retiring late

If you defer the date of your retirement you should get much better terms when you finally give up work. This is because you will be claiming the pension for less time. You get around $7\frac{1}{2}$ per cent increase for each year you have not claimed from the DSS, a bit more from a company pension.

State pension

The DSS will start to pay your pension to you five years after the due date – that is sixty-five for a woman, seventy for a man – regardless of whether or not you retire then. It doesn't come to you automatically, you do have to claim it, but if you don't and leave it until you are, say, seventy-five you won't get any more by the further deferment, so you might as well claim. Of course, if you are still working your tax bill will go up because a pension – from whatever source – is taxable income.

Company pensions

You will get better terms from a company pension if you put off claiming it – even if you are not paying into the scheme during the last few years. If a sixty-five-year-old man on £20,000 a year works on for another five years – but doesn't pay into the pension scheme or claim any pension during that time – his pension will rise from £10,000 a year to £15,000 a year at seventy.

MARRIED WOMEN AND THEIR BASIC PENSION

Married women can claim a state pension on their own National Insurance contribution record, or on the contribution record of their husband. But not both. If the claim is on their own record, the pension will be paid from their sixtieth birthday. If the claim is on their husband's record, the pension will be paid from his sixty-fifth birthday, even if the woman reaches sixty before he reaches sixty-five.

In the past, many married women opted to pay a much smaller NI contribution, aiming to collect a pension on their husband's contribution record.

This practice was stopped in April 1977 – if you hadn't opted for the 'small stamp', as it was colloquially known, by then, it was too late. However, you can change from the reduced contributions to the full rate. You should do so if:

- you will be retiring before your husband
- your husband's NI record is broken and you could build up a record of your own
- you want to claim, in the future, for Unemployment, Maternity, Sickness or Invalidity Benefit.

If you decide to change to full contributions, you cannot change back to the reduced rate again. So check with the DSS first that you will be paying the full contribution for enough years to qualify for at least part of the basic pension in your own right, before you switch.

CLAIMING YOUR PENSION

State pension

Around four months before your state retirement age, the DSS will automatically send you a pensions claim form (if they don't, write and ask for one). Even if you don't know the exact date that you are retiring, fill in the rest of the form and return it.

Your pension will be paid to you:

- monthly or quarterly into a bank or building society account
- weekly at the post office of your choice.

If you choose to be paid cash at the post office, you will receive your money on a Monday or more probably a Thursday (Tuesday if you are a widow). You can specify which day of the week you would like. For example, if your husband has a Thursday pay-day, then you can opt for a Thursday. But you must ask when you notify the DSS of your retirement day. You can't change your pay-day once you have started to receive your pension. The pension is paid from the pay-day following your retirement until the pay-day after your death.

Company pension

Your company will get in touch with you prior to your retirement to let you know the terms of your pension – how much you are entitled to and how it is going to be paid. The company may also offer you a pre-retirement course or specialist financial advice. Go along and listen to what is said – but my experience has always been to take that advice cautiously. Read up about investment yourself so that you know which of the options will best suit your financial circumstances.

COMPANY PENSIONS

Employers set up pension schemes as a perk of the job. A good scheme can attract and keep quality staff because it will offer the sort of benefits that employees want:

- a retirement pension
- a pension that may be increased, after retirement, to offset inflation
- generous treatment if you retire early through ill-health
- the option to retire early
- good widow's, widower's or dependant's benefit, if you die before or during retirement, because there is often a lump sum and a pension.

Your employer has no obligation to provide you with a pension or a pension scheme. If he starts one up, though, he has to contribute. In the past you were often forced to join the pension scheme as part of your employment contract. If you didn't want the pension, you couldn't have the job.

This has changed. You now have no obligation to join your employer's pension scheme.

The rules changed in April 1988 – you can now opt out, or not opt in, to the company pension scheme which may in the past have gone with the job. The big question is – can you do better elsewhere? Before you can answer that, you really have to understand how a company pension scheme operates.

There are two main types of company pension scheme:

- final salary
- money purchase.

Final salary

This is the more common type and covers most large and middle-sized firms that have pension schemes. It guarantees you a pension based on a percentage of your final salary.

The longer you have worked for the company, the larger will be the percentage you will get.

Most schemes work on the formula of $\frac{1}{60}$th or $\frac{1}{80}$th of final salary for every year worked or every year in the pension scheme (it depends on the rules of the scheme).

So, if you have worked for thirty years, you could get an annual pension of $\frac{30}{60}$ of your final salary. That is a half, every year.

There is usually a maximum of $\frac{40}{60}$, so even if you have worked for forty-five or even fifty years before you retire you will only get $\frac{2}{3}$ (or $\frac{40}{60}$) of your final salary, and there's a government limit to the size of your final salary for pensions purposes. At the moment it is £75,000 (rising with inflation every year). So even if you earn, say, £100,000 a year, the tax-efficient part of your pension will be limited to $\frac{40}{60}$ of £75,000 – that is £50,000.

However, this applies only to schemes set up after 14 March 1989 (the day the Budget introduced the limits) or to employees joining any scheme after 1 June 1989.

Final-salary schemes are useful:

- in times of high inflation. Wages tend to keep pace with inflation so, if your pension is based on your final salary, its value will be going up. Even if you only experienced 5 per cent inflation throughout your working life, your final salary would be more than eight times the amount of your starting salary (not counting any promotions!)
- if you are promoted, because the more you earn the larger will be your salary and your pension.

Money purchase

For many employers and employees inflation of more than 20–25 per cent a year is now a long-forgotten memory of the 1970s. So, money-purchase schemes are coming back into popularity.

With this type of scheme, the pension contributions from the company and the employee (if he contributes) are used to build up an individual fund. When the employee retires, his fund is used to buy a pension from an insurance company. Money-purchase schemes are more common in small firms running pension schemes.

Why bother?

If a company sets up a pension scheme for employees, the company has to make some contribution to it. Whether or not the employees also contribute depends on the rules of the pension scheme.

All employee contributions are tax-free – that means that no tax is deducted from the money that goes into the pension scheme. You can pay in up to 15 per cent of your gross salary every year.

How much you have to pay in depends on the rules of your particular scheme.

Very few indeed would insist on 15 per cent, though in some lines of work, such as the fire and police service, where the retirement age is low, their contribution is often over 10 per cent. At the other end of the scale, the pension scheme may be 'non-contributory' – that means that employees don't have to pay in anything at all.

The great benefit of a pension scheme is its tax efficiency. Not only are contributions to it tax-free, but the money invested in the fund is allowed to grow free of all tax, and, if you take part of your pension as a lump sum when you retire, that comes to you tax-free as well. How much you can take depends on the rules of your company pension scheme. There is now a maximum lump sum you can take tax-free. The limit is actually on the amount of earnings that can be taken into account when calculating the lump sum. If the scheme started on or after 14 March 1989, or you joined an existing scheme on or after 1 June 1989, then only the first £75,000 of your annual earnings can be used. This means that the maximum lump sum you can get out is £112,500.

All pension schemes, whether final salary or money purchase, rely on the investment managers. The better the money is invested the higher could be your pension at the end. And this is something that the individual pension members have little control over. A badly managed fund, which invested the money in shares which fell sharply over a long period, could cost you part of your pension on your retirement.

Contracted out

This is the final piece of pensions jargon that you really have to understand. A company scheme which is contracted out has removed its members from the SERPS part of the state pension scheme. Both employees and employer pay lower National Insurance contributions.

In return:

- a final salary scheme has to pay you a guaranteed minimum pension which will be at least equal to what employees would have got from SERPS. In practice most schemes provide more than SERPS
- a money-purchase scheme will have to use the money saved on the National Insurance contributions, by both employer and employee, as the guaranteed minimum contribution to the pension scheme. This will replace the SERPS pension.

Topping up your pension

If you want to make additional contributions to your pension, take out what are known as AVCs – additional voluntary contributions.

You can pay up to 15 per cent of your gross salary tax-free into a company pension scheme every year. If you pay less than that you can make up the difference, or part of the difference, through AVCs. In the past AVCs had to be regular payments but now you can pay them any way you like – regularly, spasmodically, or one-off payments.

AVCs can be paid into a company scheme or to a separate financial institution, in which case they will be known as FSAVCs – free-standing additional voluntary contributions.

You can start paying AVCs at any time in your working life, but most people leave it until the last ten to fifteen years or so because by then they usually have the mortgage under control, the children off their hands and a larger wage coming in. AVCs are a very tax-efficient way of saving for retirement.

If you put the same money into a building society account, for example, it would be taxed before it went in, and the interest would be taxed, so all the tax advantages of a pension scheme would be lost. But of course you could withdraw the money before retirement if you fancied splashing out on a new car or a glorious holiday. You can't do that with pensions savings.

Changing jobs

Until a few years ago, changing jobs could seriously damage your pension. People who changed jobs several times during their working life, paying into company schemes all along the way, could find themselves badly hit on retirement because every time they moved jobs they had to leave their pension behind. There it remained, uncherished and unchanged, until they retired.

No more. The rules have been changed so job-changers now have a choice of what to do with their pension, and the range of options irons out the gremlins that hit job-changers' pensions in the past. They can:

- get a refund of contributions
- leave their pension behind
- transfer their pension to a new scheme.

Refund of contributions

If you have worked for your employer for less than two years, you are, in most schemes, entitled to a refund of your contributions. In a good scheme you might get interest added to this money, otherwise all you will get is your money back, less tax (currently 20 per cent), and, if the scheme is contracted out, less a deduction to reinstate you in SERPS. Your employer's contribution will not be added on.

Leave your pension behind

This is known as a preserved pension. If you have been in the pension scheme more than two years (or have worked for over two years of pensionable service) your employer must allow you the option of leaving your preserved pension in the company pension scheme when you go. That means the pension scheme will preserve your pension plus an appropriate revaluation for inflation.

That increase must be at least in line with inflation every year (up to a maximum of 5 per cent a year) – although if you left before 1 January 1991, there was only an obligation for the pension (other than the guaranteed minimum pension in contracted-out schemes) to be increased from 1 January 1985.

For a final-salary scheme, if you left before 1 January 1991, your pension will be based on what you were earning when you left the company, not what your salary was elsewhere just before you retired. After that date, leavers' pensions must be revalued. In a money-purchase scheme, your pension contributions and those of your employer will continue to be invested and grow.

Transferring your pension

It is not just footballers who have transfer values – your pension does as well.

Your pension scheme will tell you the transfer value of your pension – and that money can usually be paid into:

- your new employer's scheme
- a personal pension plan
- an insurance company transfer plan.

What you cannot do is take the transfer value into your bank account and spend it! The Inland Revenue gives full tax relief on pension contributions because you are providing for your old age. It does not see a pension scheme as a tax-free piggy bank that you can empty when you feel like it.

Preserve or transfer?

When you decide to apply for a new job, your pension rights are possibly the last thing you look at. So the first time you think about your pension, in relation to your new job, is when you are offered the option of transferring your accumulated pension or preserving it.

Unfortunately there is no easy rule of thumb to tell you what to do. Each case stands on its own merits.

Your pension fund will provide you with a statement of your preserved pension, and a note of your other rights, such as death benefits, likely pension increases after retirement and so on. They will also quote a transfer value for your pension.

What you have to do is try to compare the two. Ask your new company what the pension scheme will offer you, in the way of likely pension and benefits, for the transfer value. Or get a quote from an insurance company for a Section 32 buy-out, a personal money-purchase scheme based on the transfer value invested. Try to prompt them to quote on the same basis as your preserved pension as it will be easier to compare the two if there are no differences in such areas as retirement age, continuation to a dependant on death or provision for increases in retirement.

But don't take a narrow approach to this. If you go from a scheme which pays no increases in retirement to one which pays 3 per cent increases don't compare the two by asking the new scheme for a quote based on no increases. In most cases the scheme offering increases will be better than the one which doesn't.

Although you have a right to take a transfer value, you have no right to transfer money into a pension scheme. An employer could say that he doesn't want your transfer value – though it is very unusual for a company not to allow someone to arrange a transfer value into a pension scheme.

Pensions from the past

If you have bits and pieces of pensions dotted around various firms as a result of past job changes – or even just one large lump lying somewhere – you might be able to unfreeze it. Pension schemes can refuse to let the money out if you left the scheme before 1 January 1986 though.

SELF-EMPLOYED

Until July 1988, the self-employed (or people working for a company that didn't provide them with a pension scheme) could provide a pension for themselves by taking out what was known as a Section 226 plan. That is because they were approved under Section 226 of the Income and Corporation Taxes Act of 1970!

These have now been replaced by personal pension plans but any existing Section 226 plans which are paid with regular monthly or annual premiums can continue to run.

PERSONAL PENSIONS

Anyone with earnings is entitled to have their own personal pension plan. They work like company money-purchase schemes in that you pay money into your own personal fund. When you retire, the money is used to buy an annuity from an insurance company that will pay you a regular pension for the rest of your life, though you would be able to take some of the proceeds as a tax-free lump sum.

If you buy a personal pension plan – from a bank, building society, insurance company or unit trust company – and use it to contract out of SERPS, part of your and your employer's National Insurance contributions (called the contracting-out rebate) will be paid into it. Until 1993, it will be 5.8 per cent of your earnings between the upper and lower NI limits (in 1992–3 £54–£405 week). After 1993 that percentage will drop to 4.8 per cent.

That contracting-out rebate, which will be paid into the plan after the end of the relevant tax year, is the minimum contribution that can be invested. You can pay in more – either into that plan, or another one – up to a certain percentage of your earnings in a tax year, depending on your age:

17.5 per cent of earnings for those aged	35 or less
20 per cent	36–45
25 per cent	46–50
30 per cent	51–5
35 per cent	56–60
40 per cent	61 and over

Contributions relating to earnings of over £15,000 a year won't get tax relief, so there's no point in paying in 40 per cent of a £100,000-a-year salary and expecting full tax relief. You won't get it.

If you pay more than the contracting-out rebate, when you retire the money in the fund will be used to buy an annuity which will pay you a regular pension for the rest of your life, or again you can take part as a tax-free cash sum.

To encourage people to take out their own contracted-out personal pensions the government will add an extra 2 per cent to the 5.8 per cent the DSS is already paying until 1993. You won't get this extra 2 per cent if you have been a member for more than two years of a company scheme which is 'contracted out' of SERPS. From 1993, there will be a 1 per cent bonus for anyone aged thirty or over.

With personal pension plans that are based on the minimum contracting-out rebate, extra benefits such as widow's and dependant's rights, increases after retirement, provision for early retirement due to ill-health and so on are likely to be lower. If you want these benefits to be better quality, you will have to pay higher premiums.

Types of plan

There are three main types of personal pension plan:

- deposit administration
- with-profits
- unit-linked.

Deposit administration. This is a pension fund which is ultra-safe because it grows by just having interest added to it. If you are in your last few years before retirement and want to take no risk whatsoever with your money this is for you.

With-profits. You know when you take out the plan what the absolute minimum will be that you can get when you retire. Bonuses are then added yearly and there may well be a terminal bonus at the end to top it up further. The level of bonus is determined by the investment performance of the pension fund manager and the state of the stock market throughout the life of your pension, because some of your pension fund money is invested in shares. If the market is going up, bonuses will be high, but if it is going down –

Unit-linked. Your pension contributions will be invested in the stock market so the performance of your fund will be linked to various investments. In rising markets you will do very well indeed but share prices do fall as well as rise. Unit-linked investment is more directly related to the ups and downs of shares; with-profits will give you a smoother ride.

However, if you are a long way off retirement when you start your personal pension plan, you might opt for this riskier version, but switch later.

You don't have to put all your eggs in one basket. You can take out more than one plan; spread your pension contributions about from one pension provider to another or just scheme to scheme provided you don't put more than the allowed level of your earnings (see the table opposite) into the scheme in any one year.

You can't remain in a company pension scheme and start up a personal pension plan (except in very limited cases) – though you can have a company pension scheme and free standing AVCs (see p. 326), or pay AVCs to the company scheme.

Which pensions option is best for you?

There is no clear-cut answer to this question, I am afraid, but here are some guidelines:

Does your company already have a pension scheme?

No. Then you have the choice of staying with the existing state pension (OAP plus SERPS), contracting out of SERPS through a personal pension plan that can be funded solely by the DSS SERPS contributions and/or putting more money in yourself. In many ways this is the easiest of all the pensions questions to answer. What you do depends on your age.

When you are young the DSS 5.8 per cent will buy much more than the SERPS pension you are giving up, but when you are older it will not buy nearly as much as the SERPS pension you are giving up.

Yes. Then you have the choice of staying in the company scheme or moving out into a personal pension plan. This time it is not so easy to decide because you cannot put a value on what either plan can offer you at retirement date. But you can compare the two types, as shown opposite.

Stick with your company scheme if:

- the scheme is non-contributory – you are getting something for nothing; if it is not enough you can always top it up with AVCs
- you are in ill-health – you might get good pension terms for early retirement
- you see voluntary redundancy on the horizon – you might get good early retirement terms
- you intend staying with the same company until retirement – you will get a better deal
- the scheme is money purchase and the employer pays more than the contracted-out rebate – you will get a better pension
- you value the added extras – such as dependant's rights, life insurance and so on
- you earn over £75,000 a year and joined before June 1989; you won't be affected by the new £75,000 maximum rules.

Take out a personal pension plan if:

- your company doesn't have a pension plan, isn't going to start one up or won't allow you to join

Company pension scheme v. personal pension plan		
	Company	*Personal*
Contributions	Employer has to contribute	Employer has no obligation
Retirement date	May be able to retire and get good pension if through ill-health or redundancy	If contracted out the SERPS part can only be collected at 65 or 60, the rest can be paid from age 50, but will be much reduced if taken early
Family protection	Usually provides cover for death in service, ill-health and death after retirement	Death in service cover depends entirely on amount of money in plan at time of death. Life insurance costs extra
Job-changing	*Final salary.* Can be hit by job-changing because transfer value may be lower than expected and if left behind could be based on leaving salary, not pre-retirement salary. May be a problem if under two years' pensionable service *Money purchase.* No problem on job-change because employee can take a transfer value	No problem on job-change because pension is not linked to the job
Pension level	*Final salary.* Retirement pension linked to final salary *Money purchase.* Pension linked to fund's investment performance	Pension linked to fund's investment performance

- you think you will be moving jobs within two years of joining the company you are with – otherwise, in most schemes, you will only get your contributions back from the company pension scheme
- you think you will be changing jobs a lot over the next few years – it simplifies your pension arrangements
- you are self-employed and have no Section 226 plan.

There is one other factor to bear in mind – and that is the cost of running the pension fund. It is more expensive, per person, to run a personal pension plan than a company scheme. So the administration charges will be higher on personal plans – and the salesman selling you the pension will also get a commission in most cases.

That money comes out of your pension premiums, leaving less to build up in your fund.

If a salesman is persuading you to buy a personal pension plan, ask the following questions:

1 What will my monthly contributions be?
2 What will my pension be when I retire?
3 What happens if I retire early? – five years early? – ten years early?
4 How early can I get my pension?
5 What happens if I retire early through ill-health?
6 What are the pension administration charges?
7 What is your commission?
8 What happens if I die before I retire?
9 What happens if I die after I retire?
10 Can I cancel this agreement?

Compare his answers with your company scheme by asking your company pension fund the following questions:

1 What are my monthly contributions?
2 Is my pension final salary or money purchase?
3 What will my pension be when I retire if I stay with the company?
4 What happens if I retire early? – five years early? – ten years early?
5 How early can I get my pension?
6 What happens if I retire early through ill-health?
7 What is the company contribution to the pension scheme?
8 What happens if I die before I retire?
9 What happens if I die after I retire?
10 What other benefits does the pension scheme offer?

28 Housing for the Elderly

If you are house rich, cash poor, what is the best way to raise money on your home? · Sheltered housing · Residential and nursing homes

See also

- *Chapter 6, Life Insurance*

Reaching retirement age doesn't mean you should be reaching for your bath chair. Nor should you immediately assume that you are now 'old' so you should change your style of life . . . and your style of house.

Hundreds of thousands of pensioners continue to live happily and ably in the home they have had since marriage. But thousands of others don't.

It may be that they need somewhere smaller, without stairs, with a warden or near their children. It may just be that they can no longer afford the upkeep on their home. It is reckoned that a sixty-five year old will spend £20,000 on bills and maintenance over the rest of his or her life, and most old people do not have that sort of cash. The problem is that the capital is all locked up in the bricks and mortar. They are house rich, but cash poor.

RAISING MONEY ON YOUR PRESENT HOME

What you need, if you are in that category, is the key to unlock the cash without having to leave your home to do it. And that key comes in three different forms:

- home income plans
- reversion policies
- interest-only mortgages.

There are two major drawbacks:

- You have to be quite old to qualify. A single person would have to be at least sixty-nine, and a married couple would need a combined age of at least 145 years (though more usually 150 years) to qualify – but to get real money out of your house you would have to be older than that.

- If you are on Income Support, any income you get would be deducted from your benefit and if you get a lump sum taking your savings over £3,000 your Income Support could be reduced. If your savings go over £8,000 it will be withdrawn altogether. Losing your Income Support means you lose other rights, such as your free prescriptions and home help, and so on, and if you are receiving housing benefits these will be reduced or even lost, so make sure any increase in income from your house is worth it.

Home income plan

A home income or mortgage annuity plan is the most popular choice for most elderly home owners – probably because it allows them to retain ownership of their home. And it works like this.

Mrs Brown is a seventy-five-year-old widow living in a two-bedroom semi. She has a state pension and a small company pension from her late husband's job. She doesn't want to move house but feels she might have to because she can't afford to get the roof repaired, and the fuel bills are too high for her income.

With a home income plan Mrs Brown raised £30,000 on her £40,000 house. She kept £1,000 to repair the roof. The other £29,000 she used to buy an annuity. She gets £3,540 a year, from which she has to pay tax and the interest on her loan. That leaves her with £2,900 a year net, or £242 a month for the rest of her life, no matter how long she lives. If Mrs Brown's income is little more than her state pension she won't be a taxpayer. The plan can then be drawn up on what is known as a non-tax basis – she would get £1,571 a year, which is £131 a month.

When she dies, her daughter can sell the house, pay off the £30,000 loan and the rest of the money will be hers. If Mrs Brown lives a few more years, the house should rise in value and any increase will go, on her death, to her daughter.

What it is. You borrow a lump sum using your home as security. That lump sum is used to buy an annuity which will be large enough to pay the interest on the loan, and provide you with a monthly income for the rest of your life. On your death – or the death of the last survivor if you are taking out the plan as a couple – the house is sold and the loan paid off. Anything left over goes into your estate to be willed to your heirs.

You cannot borrow the full value of the house. Most lenders will only advance up to 70–80 per cent, with a top limit of £30,000 – that's the MIRAS limit for tax relief on the interest. It does mean, though, that there will always be something left over to pass on to your heirs.

You do have to pay some costs when you take out a plan.

There would be a valuation survey – around £80 on a £40,000 house – and there will be solicitor's charges. These are often refunded in part or in full if you go ahead with the plan.

Advantages. The home remains your property because all you have done is raise a loan on it. If the house increases in value over the years, that increase is yours because the size of the loan doesn't increase.

Drawbacks. The amount you can raise on your home is limited and the money has to be paid into an annuity from which you will get an income. That may not suit you if you need £5,000 up front to put a new roof on your home – though some schemes will pay you up to 10 per cent of the loan as a cash lump sum. The other great drawback of using the money to buy an annuity is that if you die the day after it is taken out, the income from the annuity is lost immediately but the full loan that was borrowed still has to be paid back. You can insure against that happening. If you die in the first three or four years, your heirs would only repay a proportion of the lump sum back, but that means, your monthly lifetime payments from the annuity will be smaller.

Before going ahead. Get plenty of annuity quotes to see who is offering the best deal at the time. Check whether the interest rate is fixed – that means it will stay the same for the rest of your life no matter what happens to interest rates in general. If it is variable, it will go up and down with interest rates so the amount you are paid every month from the annuity will vary as interest rates change, and few retired people want that.

If you think you might move home at some stage, find out how flexible the plan is to cope with that contingency.

How much should you get. If your house is valued at £40,000 or more and you want the maximum possible:

- a 73-year-old man could get £1,729 net per year (non-tax basis £2,107)

- a 75-year-old woman could get £1,471 net per year (non-tax basis £1,814)
- a couple (husband 78, wife 75) could get £1,057 net per year (non-tax basis £1,368).

If you decided to insure the loan so that if you die in the first year only 20 per cent of the lump sum has to be paid back, 40 per cent in the second year, 60 per cent in the third year and 80 per cent in the fourth year, then:

- a 73-year-old man could get £1,504 net per year (non-tax basis £1,808)
- a 75-year-old woman could get £1,286 net per year (non-tax basis £1,567)
- a couple (husband 78, wife 75) could get £1,037 (non-tax basis £1,342).

The actual percentages and costs change depending on which company you take your home income plan out with.

Reversion schemes

Mrs Brown, a seventy-five-year-old widow in her two-bedroom semi, which is worth £40,000, decides instead to opt for a reversion scheme. She buys an annuity with the money raised and gets £2,577 gross a year. Under this plan she only has to pay tax – there is no interest to pay because there is no loan, so her net income is £2,254 a year or £188 a month for the rest of her life.

When she dies, the house goes to the reversion company, and her daughter gets nothing.

What it is. A reversion scheme also allows you to raise money on your home – but this time you lose the ownership in exchange for life-long tenancy. Normally you would get between a third and half of the value of your house, depending on your age – the younger you are, the less you get. Men get more than women because of their shorter life expectancy. I always think smokers should get more too – but they don't.

The money is used to buy an annuity. You remain in the house, responsible for the upkeep and the running costs, and paying a peppercorn rent of around £1 a month or so. On your death – or the death of the survivor in the case of a couple taking out the scheme together – the house would go to the reversion company.

There are variations on that basic concept. Some schemes allow partial sales so that your heirs will be able to inherit the part of the property that you retain. Others offer the owners a share in rising house prices. They will give you a lower annuity when you take out the plan but in return they pool all the property values and as the value of the pool rises, so your income rises by the same amount (up to a maximum of 12 per cent a year). Of course in time of depressed and falling property prices, your income will fall but it is possible to get a guarantee that the income will never fall below the opening level. A recently introduced scheme works on a different basis. You sell a slice of your house – for example 20 per cent – and this provides you with a guaranteed income for five years. At the end of this period, the income stops but you can then sell another slice for another five years of income – and so on until you reach eighty-five, when the income continues for life. Using this type of plan you are more likely to benefit from a rising property market, and most people would sell less of their house, so something would be left for the heirs.

How the plans compare:
annual income received on a house valued at £40,000

	Reversion plans		*Home income plans*	
	fixed income	*income linked to house values*	*tax payers*	*non-tax payers*
Man aged 73	£2,600	£2,177	£1,729	£2,107
Woman aged 75	£2,254	£1,806	£1,471	£1,814
Couple (man 78, woman 75)	£1,838	£1,343	£1,057	£1,368

The figures given here are for comparison only. What you actually get will depend on your age, personal tax rate and the value of your home.
(The home income plan would only advance £30,000 so there would be £10,000 left in the house.)

Advantages. Reversion schemes tend to suit the elderly who have no children or near relatives to leave their house to, or whose heirs are already well-off. If they have a valuable house they will be able to raise more money this way because there is no top limit of £30,000 and none of the monthly income has to be used to repay interest. They can also take the money in a lump sum, which could be more useful to some people than a monthly income.

Disadvantages. The owner loses his home and in most cases has nothing to pass on to children or heirs. There is generally no protection against inflation because the owner receives no benefit from any increase in the value of the property after the plan has been taken out.

Interest-only mortgages

These are like home income plans but suit people who are not old enough to qualify. They work, as their name suggests, like an ordinary mortgage on which you only pay interest. At the end of the term, or when the holder sells the home, then the capital is repaid.

Warning
Don't use schemes that include roll-up loans, or investment bonds. Initially they may seem a better deal, as you appear to get a larger income from your house. But if interest rates or investment decisions turn against you, you may end up having to sell up to pay bills.

Watch-dog body

Some of the companies which provide home income plans have recently formed a watch-dog group to safeguard people thinking of taking out such plans. In their code of practice are clauses to ensure that you and your lawyer are absolutely sure you know what is happening and what you are taking on and giving up to get the monthly income.

You can get a copy of the Safe Home Income Plan (SHIP) code of practice from SHIP Campaign, 374–8 Ewell Road, Surbiton, Surrey KT6 7BB.

Who does what

The companies which specialize in home income plans are:

- Hinton & Wild (Home Plans) Ltd, 374–8 Ewell Road, Surbiton, Surrey KT6 7BB (tel. 081 390 8166)
- Carlyle Life Assurance Co. Ltd, 21 Windsor Place, Cardiff CF1 3BY (tel. 0222 371726)
- Allchurches Life Assurance Ltd, Beaufort House, Brunswick Road, Gloucester GL1 1JZ (tel. 0452 26265).

The companies which specialize in reversion schemes are:

- Stalwart Assurance, Stalwart House, 142 South Street, Dorking, Surrey RH4 2EU (tel. 0306 876581)
- Carlyle Life Assurance Co. Ltd (see above)
- Hinton & Wild (see above)
- Home & Capital Trust Ltd, 31 Goldington Road, Bedford MK40 3LH (tel. 0234 340511).

SHELTERED HOUSING

Sheltered housing, sometimes known as warden-assisted or warden-controlled housing, is becoming increasingly popular. It is often a half-way house for able-bodied elderly people who can no longer cope with their own house but don't want to go into a home.

In the public sector, council-, church- or charity-run schemes are difficult to get into in many places.

In the private sector, developers, some councils and housing associations are now building sheltered accommodation for sale to the elderly. Apart from the cost of the flat, you would also have to pay your own council tax and running costs, and a maintenance charge to the management company.

When you die, or want to move, the flat is put up for sale just like any other property – though you could only sell it to someone who fulfils the sheltered-housing criteria. It would be no good if your top offer came from a thirty-five-year-old.

RESIDENTIAL AND NURSING HOMES

If you move into a private nursing home, you will know what the charges are at the outset, and whether you can afford them.

The alternative is a local authority residential home. How much you pay depends on your financial circumstances.

Each local authority home has a standard weekly charge. If you can pay it in full, good and well, if not then you will be assessed on how much you can afford.

Your income and capital will be taken into account – and if you own your house that will be included in your capital. Your house will not be included if your spouse or a child under sixteen still lives in it, or if it is 'considered appropriate' not to include it, perhaps because your principal financial supporter still lives there and has nowhere else to go.

When you sell your home, your capital will be assessed and you will pay 25p a week for every £50 of capital you have.

So, if your capital amounted to £13,800 you would be charged £69 a week (that is £13,800 ÷ £50 × 25p). On top of that you would have to hand over any income you had coming in – say your state pension.

Once the local authority has worked out how much you can afford to pay, you will pay that every week until your capital is reduced to below £1,500.

If you have less coming in per week than the minimum charge – which is usually equal to the state pension – then the DSS will make up the difference and allow you weekly pocket money through Income Support.

29 Making a Will

Using a solicitor or doing it yourself · Cohabiting and divorce · Cutting someone out

See also

- *Chapter 11, Inheritance Tax*

They say there's a toast at every lawyer's dinner 'To the people who don't write a will.' Lawyers make far more from sorting out bad wills, or the affairs of people who haven't made wills, than they ever do from helping you to write one.

Everyone over the age of eighteen should have a will – no matter how little you have to leave. If you die intestate (that is without a will) it causes a lot of trouble and expense to the people you leave behind. Your estate (that is everything you own) will be divided up according to the law rather than according to what your wishes might have been. And only very close relatives can benefit.

You can draw up a will in one of two ways:

- by writing it yourself
- with the help of a solicitor or bank.

Unless your affairs are very straightforward you would be better getting professional help. But you will still cut your legal bill considerably if you go along to see your lawyer knowing exactly who you want to look after your estate when you die, and how much you want to leave to whom.

D-I-Y

If you decide to write your own will you can either use a clean sheet of paper or you can buy a printed will form from a stationer or newsagent. A will form gives you some guidance on writing your will and costs around £5.

The main rule that you must obey is to see that your will is signed and properly witnessed (the law in Scotland is different on this point).

After you have written your will, you must sign it in front of two other people, the witnesses. They are witnessing your signature, not the will, so they don't have to read the contents. After they see you sign, they together must sign in your presence and state their name, address and occupation. Witnesses must not be beneficiaries – that is, they must not be left anything by you in the will – nor must they be married to beneficiaries.

The main point to remember when you are writing your own will is that you have to be very clear about what you are leaving to whom. If it is at all complicated, don't do it yourself.

In Scotland. Although the general rules of writing your will are the same, there is one major difference:

- If you hand-write your will – it is known as a holograph – you don't need to have the signature witnessed. But you can, and most people do, get your signature witnessed by two non-beneficiaries as in England.

Warning

I do not think it is a good idea to write your own will because:

- you run the risk of getting it wrong and ending up with results you never intended
- you might make an essentially simple situation unnecessarily complicated by, for example, overvaluing the money you have coming from an insurance policy
- you have no experience of what is practical in a will and might not foresee problems
- in Scotland, you might not know about legal or prior rights that can be claimed anyway. What you write might conflict with them; for example, you might try to cut a wife or child out of your will and this cannot be done.

Solicitors

Going to a lawyer to have your will drawn up need not be expensive. How much it costs will depend on how much time the lawyer spends with you finding out how you want to pass your money on, and writing the will for you to sign.

Ask in advance and you should find that a simple will costs between £20 and £50 – a more complicated will could cost anything over £100 or so. What you pay depends on how much of the lawyer's time you take up so it is difficult to be specific on costs.

One rule of thumb is that it will cost the same as it does to get your car serviced. If you are rich enough to own a Rolls Royce, your financial affairs will no doubt be much more complicated than the couple who own a Ford Escort, and so more expensive to leave behind.

Don't go to see a lawyer without a very clear idea of what you want to leave to whom. The lawyer's office is not the place to ponder on who should benefit financially from your death – remember you will be charged according to how much of your lawyer's time you take up.

Whether you are writing your own will or going to a lawyer, fill in the will plan on p. 348 so that you have a clear idea, before the will is written, of exactly what should be in it.

For a will to be valid it has to be written – by hand, typewriter or whatever – signed, dated and the signature witnessed. It can be written on anything from an eggshell to the side of a house, providing it is legally correct – though I would recommend using paper!

But don't try leaving a video to explain your wishes. It may be nice for the family but it is not legal because it can be edited.

WRITING A WILL

When writing a will, you should appoint one or two executors. They are the people who will take charge of your affairs and 'execute' your instructions (that is, do what your will tells them to). Your executor can be your spouse, friends, a lawyer, a banker – indeed anyone. But if you appoint a professional (a lawyer or banker) they will charge for the work they do.

The beneficiaries of your will are the people to whom you are leaving money: either specific items such as your jewellery or stamp collection, or money, or your house, etc. If you are writing your own will make sure it is crystal clear who the person is and what you are leaving. Saying 'I leave £100 to my next-door neighbours' will not do. It must read 'I leave £100 each to John and Elsie Smith who currently live at 17 High Road, Liverpool.'

Tip
If you can't afford a lawyer, here's a tip. Write your will yourself and then take it or send it to your local CAB. If they have a legal clinic their lawyer might check it over for you and advise on any mistakes you have made. It won't be as legally watertight as getting a lawyer to write your will for you, but it is a lot better than not having it checked at all.

Will plan

1 The name and address of the testator – you.
2 The name and address of the executor or executors. Those are the people who will have to carry out the instructions that you leave in your will. You can choose anyone you like – but ask them first because they may not want the job. You are well advised not to choose someone older than you for this job because if they die first you will have to find a replacement!
   Many people choose a close friend or relative and either the lawyer or the bank as their executors. If the will is complicated it is a good idea to use a professional as one of the officials. A bank's charges are based on the value of the estate, not the time taken to sort out your affairs. Even if your affairs are simple but your house is valuable your dependants will have a large bill to pay the bank. A lawyer charges for his time so if your affairs are simple the costs will be much cheaper.
3 The main beneficiaries – that is the people who will benefit most from your estate: your spouse, children, or close relatives who will get your money, your house, your car, and so on. You should know either what sum of money or specific asset they are going to get, or what percentage of your estate they will get.
4 Your estate – write down exactly what you own in the way of house, car, shares, life insurance policies, building society or bank accounts, National Savings certificates or premium bonds, and any debts you have, such as your mortgage or any bank loans or HP agreements. That way you will be able to make sure nothing is left out.
5 Bequests – this is the paragraph that deals with the little things you want to leave. Perhaps it is your jewellery to your daughter, or £100 to the golf club, or your stamp collection to the little boy next door. Be very clear just who is getting what here.
6 Husband and wife dying together – in these days of car and air travel

and the increased likelihood of a couple being involved in the same accident, wills tend to include a clause as to what should happen to the estate if the husband and wife die together, or within thirty days of each other. Although you and your spouse are making separate wills you should know what you would like done with the money in the event of a joint death. If it is to go to your children – and they are under eighteen – the money will have to be left in trust. You should name the person or people you would like as trustee(s). Again you can choose a friend or relative and/or your lawyer or bank.

7 The family dying together – if you and your spouse intend leaving the bulk of your estate to each other, or to the children in the event of you both dying together, you should also leave instructions as to what should happen if you all die in the same accident. Otherwise your estate will be distributed, according to a laid-down formula, among your close relatives. You may not want this, so leave a clause outlining what you would prefer.

If you have children you should also leave a letter outlining your wishes as to what should happen to them in the event of the death of the parents. You can appoint the guardians (who can also be appointed in the will) and offer guidance as to who the children should live with. Your wishes will be taken into account, but the decision will be in the hands of the social services department and the courts as to what happens to them. They will decide what would be best, under the circumstances, for the children.

Co-habitation and divorce

If you are living with someone but not married it is doubly important to have a will. Unless your partner can prove that they were wholly or partly reliant on you for money they will not get anything out of your estate and the cash will go to your relatives. Not much fun if you have been living together as man and wife for years.

If you get divorced your ex-partner is automatically cut out of your will unless you state that you want them to remain in. But the share of your estate that they would have got could end up with the wrong people unless you rewrite your will. And, in England and Wales, marriage automatically revokes any former will, so, if you remarry after your divorce, write a new will.

1. Name and address of testator

2. Name and address of executors

3. Main beneficiaries

4. Your estate

5. Bequests

6. If husband and wife die together . . .

7. If whole family dies together . . .

Will plan

Cutting someone out

If you want to use your will to pay back all the knocks and insults you have taken over the years, go ahead and do so. You can leave your money to whom you choose with certain exceptions:

- in England and Wales you cannot cut out anyone who is financially dependent on you. If they challenge your will, a court will decide how much they should get
- in Scotland a wife and children have an automatic right to a certain percentage of your estate. Even if you have not spoken to, say, one of your sons for twenty years, he is still entitled to his share.

Where should you write your will?

Some people think that just because they move from Scotland to England, or the other way round, they have to make a new will. Not so. You should make a will in the country in which you regard yourself as having your permanent home. So if you move to England but intend going back to Scotland at some stage, don't make a new will in England when you get there. And have only one will.

You will find my will

My employer is

Pension details

House and mortgage details

Savings accounts

Shares, unit trusts, etc.

Car

Insurance details

Life assurance

Tax office

Other important documents

In the event of my death please contact

Life plan

LIFE PLAN

You should also draw up a life plan to leave with your will, or a copy of your will.

There is no point in being meticulous about who gets what if they can't find all your worldly wealth.

Your life plan should cover everything you own and make life easy for your executors. It should include:

- information on the house, whether or not you have a mortgage and a mortgage protection plan and which company it is with
- your job, detailing the name and address of your employer, whether

or not you are in a company pension plan and any other relevant details such as share option schemes

- your savings accounts, giving names and numbers of all the building society, bank, National Savings and TESSA accounts you have and the numbers of all your plastic-card accounts
- your car, whether it belongs to you or the firm and whether there is a bank loan outstanding
- anything else that is relevant, such as life insurance policies, personal pension plans, any shares you might own or PEP schemes you are in, or just anything in the house that is valuable and might be overlooked.

And make sure that someone knows where you keep all your personal papers and files.

30 Professional Money Advisers

Where and when to get professional help – what it costs · What are your rights

Everyone, including myself, needs professional money advice at some stage.

I wouldn't be able to cope with the fine print of inheritance tax planning without turning to an expert for help. Others may need professional help when it comes to investing for school fees or opting for a personal pension plan. Some people may just want a bit of a pointer on which type of life insurance plan to buy.

Even if you have read this book from cover to cover you may not be able to cope with all the intricacies of your financial affairs. But you should have something that is much more important – enough knowledge of money affairs to ask the right questions and weigh up the answers.

WHEN TO GET PROFESSIONAL ADVICE

It is difficult to know exactly when you should turn to the professionals for help. No matter how much you know about a particular area, the right professional will always know more because he is a specialist who is dealing with other people's cases and problems, and so has the advantage of practical experience.

When you feel you can't find the right answer to your problem, or think it would take you too long to do so, that is when to turn to the expert.

You may know that you could get cheaper car insurance, but don't fancy ringing round twenty companies for quotes, so ask an insurance broker for help. They will be able to use their computer to winkle out the best quotes, and, if they know their stuff, they'll be aware of which companies offer the best deals for your circumstances.

Whatever professional financial advice you want, make it very clear to the professional exactly what you are asking. Think about your problem before going along to see them, or inviting them to see you, so that you can ask clear questions. There is no point in making a woolly statement such as 'I want to invest some money.' You should have thought out some basic parameters, such as how much you want to invest, for how long a period and what sort of risk you are prepared to take. Even with that little bit of advance planning you could stop yourself being sucked into something you don't want.

WHAT DO THEY COST?

No one works for nothing – not me, not you, not the professionals. So don't think that, just because you are not presented with a bill at the end, you are not paying for the service.

If you are not asked for a fee, the professional is getting paid elsewhere, usually by commission. So, in the end, it is you, the customer, who is footing the bill.

Life insurance, unit trusts, pensions, endowment plans, are all commission-based. If the salesman doesn't tell you how much he is getting, ask him. Under the Financial Services Act, he is obliged to tell you. Don't be afraid, ask him. You wouldn't sit down in a restaurant and choose your meal from a menu that didn't have prices on it, so why should you choose something as vital as a pension or life insurance policy without knowing how much it is costing you. After all, you might be unlucky enough to be sold the pension by a salesman wanting the largest commission, rather than the best deal for you.

If you know the rates of commission on all the pensions that might suit you, you will be able to see at a glance if the one he is pushing hardest is the one paying the highest commission. Of course it might still be the best pension for you – but it might not!

Some other professionals, such as accountants and solicitors, charge by the hour. So if you buy a commission-based policy, such as life insurance or a pension, through them, ask for the commission to be deducted from your bill. Paying once is bad enough, but paying twice . . .

Stockbroker's commission is added on to the cost of the shares you buy and sell. The advice they give you will be free – but the smaller your portfolio, the less of their time you are likely to get.

WHO ARE THEY?

Financial advisers

A financial adviser used to be anyone who tried to sell you anything from a unit trust to a life insurance policy. It could be a well-qualified tax accountant or a young lad with a clipboard conducting a 'market research survey' outside the station.

No more. Under the Financial Services Act, anyone advising on investments (including life insurance and pensions) has to be authorized. Most of the independent intermediaries will be registered under:

- FIMBRA – the Financial Intermediaries, Managers and Brokers Regulatory Association. If they are members they will have the FIMBRA logo on their letterhead – so look out for it. FIMBRA is based at Hertsmere House, Hertsmere Road, London E14 4AB (tel. 071 538 8860).
- IMRO – the Investment Managers Regulatory Organization. IMRO is based at Broadwalk House, 5 Appold Street, London EC2A 2LL (tel. 071 628 6022).
- LAUTRO – the Life Assurance and Unit Trust Regulatory Organization. It covers most of the insurance companies, unit trusts and friendly societies. LAUTRO is based at Centre Point, 103 New Oxford Street, London WC1A 1PT (tel. 071 379 0444).
- SIB – the Securities and Investments Board. It is the private body to which the government has delegated its powers to regulate the financial services industry. SIB is based at Gavrelle House, 2–14 Bunhill Row, London EC1Y 8RA (tel. 071 638 1240).

Your adviser might also be a member of the Corporation of Insurance and Financial Advisers or the Institute of Insurance Consultants.

Don't choose an adviser just because he drives a BMW or has a suite of offices in the poshest part of town. The trappings of wealth do not pass muster as credentials.

Independent financial adviser

You should be able to get advice on a wide range of financial products from an independent financial adviser – and he or she has to give you what is known as 'best advice'. That means the most suitable product available from the market as a whole for your needs and financial circumstances. That is now the law. So an adviser is no longer allowed to recommend the products just because they pay the best commission.

The adviser must also 'know the customer', so unless he asks you relevant questions about your financial affairs – how much you are investing, what your tax position is, how much risk you want to take, how long you want your money tied up and so on – use someone else. (If your adviser is only buying or selling for you, not offering advice, this rule does not apply.)

Tip

Never make your cheque payable to the financial adviser. Always write it out to the company whose product you are buying.

So if Jack Adviser sells you a Big Company life insurance policy, make the cheque payable to Big Company.

One-company salesman

Under the Financial Services Act, anyone selling life insurance or unit trust products has to be tied to one company or group, or register as an independent intermediary. If they are actually a company representative, or belong to a firm of investment advisers who only sell one company's products, they must make this clear to the client.

Although they are only selling one company's products they must still find the policy that will suit the customer's needs best from within the company's products – if there isn't one, they must admit that!

A company salesman doesn't have to tell you how much commission he is making on a product unless you ask – so ask. Commission rates vary from company to company; they also vary from product to product. A life insurance policy pays more than unit trusts, so bear that in mind.

Insurance brokers

If your adviser calls himself an insurance broker, he must be registered with the Insurance Brokers Registration Council. Calling yourself an insurance broker and not being a member can lead to a fine of up to £1,000; he might also be a member of FIMBRA.

You won't have to pay a fee to an insurance broker – they make their living by receiving commission.

They get paid in one of two ways by the companies whose products they sell:

- on a standard scale
- more than the standard scale, sometimes or always.

If the commission on the product you are buying comes under the first heading, you will be told by the broker that the commission is on the set scale.

If it comes under the second, you must be told how much is being paid, though it will be expressed as a percentage of your investment.

> **Tip**
> If your affairs are not too complicated and you need some tax advice, try your local PAYE office. There are special Inland Revenue tax inspectors there to help you sort out your problems – and, best of all, they don't charge anything at all.

Always remember that even if you go to a professional adviser for help, you don't have to take their advice. Ask plenty of questions, check some of their answers, use this book to weigh up the quality of the advice and then decide whether or not to go ahead.

After all, it is your money and in the end it is only you who has your real financial interest at heart.

Ten years ago I was pestered by a salesman into buying a paint treatment for the outside of the house that in the end was not the right product for my type of home. It cost me several hundred pounds and a court case to cancel an agreement I should never have signed – but it saved me thousands because I have never, ever signed anything, since that day, without allowing a night to pass so that I could 'sleep' on it. It hasn't lost me a deal, but it has saved me from a few.

Any salesman, financial or otherwise, that offers you a genuine deal won't try to rush it through. If it is such a good deal that you have to sign there and then and it can't wait the twelve hours – there is something wrong with it. The Financial Services Act allows you a 'cooling off' period for life insurance policies and unit trusts, unless you have an agreement to the contrary.

Most salesmen won't try to sell you something you don't need or want, and if they do, check through this book so that you can read up about the product and decide if it suits your financial circumstances.

I hope that my book will prove to be your best financial adviser.

Index